Nikolaus Mohr
Gerhard P. Thomas

Interactive
Broadband Media

Edition Accenture

Edited by Thomas Herbst, Managing Partner Accenture

The series is tailored to meet the information needs of top executives in high-tech, telecommunications and media enterprises. It provides excellent strategy knowledge with target orientation and specialized capabilities. The possible solutions also contain advice on how to implement the required technology and processes in organisations.

Including strategic functions such as shareholder value creation, organisation & reporting, and customer relationship management, the books provide you with top management knowledge and analyse business potential and business models in areas such as media markets, mobile commerce and application service provision.

For more information on coming titles visit our homepage
www.vieweg.de

Vieweg

Nikolaus Mohr
Gerhard P. Thomas

Interactive Broadband Media

A Guide for a Successful Take-Off

With contributions by
Ingo Draheim, Donata Hopfen, Matthias
Hülsmann, Markus Karras, Peter Weidermann,
Richard Werner and Randolf Woehrl

vieweg

Die Deutsche Bibliothek – CIP-Cataloguing-in-Publication-Data
A catalogue record for this publication is available from Die Deutsche Bibliothek
http://www.ddb.de

1st Edition August 2001

Vieweg is a company in the specialist publishing group BertelsmannSpringer.
www.vieweg.de
vieweg@bertelsmann.de

Printing and binding: Lengericher Handelsdruckerei, Lengerich
Printed on acid-free paper
Printed in Germany

ISBN 3-528-05781-5

Foreword

We are currently in the midst of the "digital revolution". Compared to the industrial revolution, which raised the living conditions of the majority of the world's population to a new level, the consequences of the digital change appear to be even more fundamental and far-reaching. Although the past decades have already seen radical change with regard to the introduction and spreading of PCs and the Internet, no end to the technical developments that have a major impact on our lives is yet in sight.

The media industry is particularly affected by these technological changes. From individual formats and innovative services to new methods of distributing content and entirely new possibilities for customer relations, all areas are directly affected by the changeover from analogue to digital. Convergence is the key word here - the merging of telecommunications, computer technology and media & entertainment is creating new markets. The development of interactive, digital broadband media is becoming a key issue for the whole industry.

As is probably the case for any important historical development that stretches over a period of time, the road to progressive digitalisation is not always smooth. The current "slowdown" of the New Economy - combined with a series of corporate failures, irreconcilable financing difficulties, and even ideas that are drying out - constitutes just one of the setbacks. Other aspects of the digital revolution, too, have undergone a phase of sobering up following the initial euphoria. One example of this is the "Full Service Network" developed by Time Warner in 1996.

However, these frustrating setbacks cease to appear quite so serious when viewed in the perspective of the positive, long-term overall trend towards a digital world. Rather than expecting them to lead to a fatal economic collapse, companies should consider them to be a sign of a healthy purge. The digital revolution continues to progress, giving way to new, future-oriented possibilities and opportunities for all media users.

This book discusses visionary ideas and realistically sums up the situation. After beginning by sketching the developing interactive broadband market, it continues to systematically analyse various possible solutions and successful business models for media enterprises faced with the problem of realigning themselves and remaining competitive under difficult circumstances. Here, adopting a new, customer-oriented approach is just one important step along the road to the digital revolution.

Frankfurt, August 2001

Gerhard P. Thomas   Dr. Nikolaus Mohr

Table of Contents

Table of Charts

Interactive broadband services are set to revolutionise the entire media landscape, but from a business perspective it is still unclear how companies will be successful in this attractive new market.

1   Interactive Broadband Evolution

Interactive broadband[1] is becoming reality in Europe, and related companies are changing and directing their activities towards this evolution. Convergence is the recent key word that implies a digital revolution merging fixed and mobile telephony, PC-Internet, broadcast, digital and interactive TV to form one single, integrated communications platform. Distribution channels are being digitised, developed and combined in a new manner. Voice, sound, pictures, mobile communications and the Internet are growing and merging into one. Broadband TV, the Internet, e-commerce, WAP and UMTS are the enablers of this evolution: movies will be loaded from the Internet in all parts of the world to all kinds of devices, television programs are being shown on the PC, shopping is done by pushing a button on the remote control, and photos or videos are sent to mobiles. Interactive broadband services allow the user to fully interact not only via PC, but also via mobile phone and TV. Programs will be completely personalised, guided by Electronic Program Guides (EPGs).

---

[1] According to the International Telecommunication Union (ITU), broadband is defined as a transmission system providing multiple channels over a communication medium such as Hybrid Fibre Coax (HFC) cable, phone line, wireless or satellite at a speed of at least 1.544Mbit/s. Here, we follow the general industry practice and consider ADSL to be a broadband technology although it often has lower bit rates.

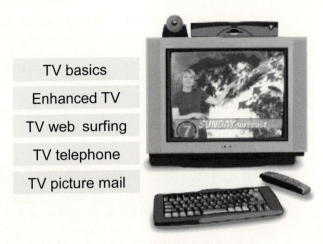

TV basics

Enhanced TV

TV web surfing

TV telephone

TV picture mail

Chart 1: Interactive Broadband Television

This appealing vision will have widespread influence on how we live and work, thereby fascinating many of us - not only the technology-enthusiasts. From a consumer standpoint, interactive broadband services will offer a richer experience, combining emotional elements of television with rational elements of the Internet world. Looking at today's media companies, the future described is coming up with a number of interesting business opportunities. Interactive services will drive the need for new, interactive forms of content that really exploit the full potential of the new medium. New forms of storytelling will be developed, creating new markets for content producers. The content aggregation business will also face serious changes when trying to offer the diverse forms of content over the complete range of different access devices. New ways of addressing the consumer with the most relevant content more directly will be found by leveraging a new quality of customer insight. Advertisers will have rich opportunities for sophisticated targeting and will be able to offer more intelligent, less frustrating forms of advertising with higher efficiency. Other companies, entering the interactive broadband arena from a more technical perspective, see huge business opportunities: interactive broadband services generate demand for high-bandwidth networks with sufficient capacity to deliver individual video-on-demand services simultaneously to every household. In addition, the evolution of interactive broadband services affects software and hardware manufacturers, who will benefit from the mass market demand for new devices. Overall, forecasts expect interactive broadband services to grow into a $37 billion market in 2003/4 in Western Europe, where at least one third is expected as *new* revenues.

With the first approaches towards interactive services in mind, some progress has clearly been made: in the 1950's, CBS aired the first truly interactive television program "Winky Dink and You", in which children were invited to draw with crayon on special plastic sheets attached to the TV screen. (The program was cancelled due to parents' complaints that children didn't use the plastic sheets, rather drawing on the screen directly.) Nowadays, the Internet revolution has paved the way for true interactivity, already including more entertaining elements such as Web TV. Initial interactive television services are already implemented, for example, in the UK, Spain and in test environments in Germany. However, a number of severe complications are yet to be overcome before the vision of "access - anytime, anywhere" can come true.

A technical evolution is needed as the necessary enabler for these interactive services. Without significant development in the technical area, no interactivity can be offered. Two different areas need to be further developed. First of all, the focus is on access technologies (UMTS, cable, DSL, etc.). Here, the two challenges are sufficient bandwidth and the integration of a backward channel. Secondly, this means technology development for using interactive content. This includes set-top boxes and other devices such as cableless keyboards at the end-customer side, as well as digital content services (DCS) including capabilities for saving, aggregating and managing the huge amounts of data necessary for interactive broadband service players.

From a marketing perspective, the most important issue is the commercialisation of interactive broadband services. This has to be focused on content and service offerings and not on access technology, since the different kinds of access technologies do not make any difference in the eyes of the consumer. Once broadband access is available, the focus has to be put on content and therewith content becomes king. Also, the networking of content has to be considered, because all the success in this business depends on the quality, quantity and distribution of content. For maximum appreciation of these trends, co-operation between the different players has to be organised and networks have to be established.

Finally, there is the business perspective. The question is how to refinance the huge up-front investments necessary. Many diverse companies are making enormous investments from different sides, trying to get a foot into the door by engaging themselves in the interactive broadband realm at an early stage in order to benefit from a quick time-to-market and to establish a competitive advantage as early as possible. Media companies and retailers, for instance, are opening up broadband-specific portals and telecommunications companies are providing future-oriented access technologies. In the German market, relevant practical examples out of today's business include companies like Klesch & Company Limited and NTL, which invested the tremendous amount of DM 6 billion to buy Hessennet - the German cable operator in the Land Hessen - and Callahan, which bought the majority of the TV network in Baden Württemberg for approximately DM 3.8 billion, already owning the majority of the cable network in Nordrhein-Westfalen. As a last example, the Klesch Liberty Media group is planning to invest DM 10 billion into the six remaining regional cable companies in Germany. In addition to these investments in infrastructure, other significant costs arise from buying necessary content and the marketing efforts needed in order to build up a completely new mass market, where merely creating brand awareness will not be enough.

When thinking about refinancing, the first idea is to let consumers pay for the content and services they use. The emerging problem in this context is that, depending on the local cultures and habits, customers are currently not willing to pay any price for access or content. In the Internet world, content has basically always been given away for free, very similar to the more than 30 German free-TV channels, leading to the phenomenon that today's Internet and tomorrow's broadband users are not willing to pay in any case for the content and services they request. A recent example that affirms this behaviour is Steven King's Internet book "The Plant", which has been published exclusively on the Internet, chapter by chapter. This book was not continued, due to the fact that too few readers paid the requested amount of money for each downloaded chapter. Another example is Premiere World, the German pay-TV channel, not reaching the critical mass of customers (currently 2.5 million) for break even, because people are not willing to pay for TV.

The attractiveness of this new market on the one hand and the problems of realisation and making business on the other hand pose one decisive question with new insistence: *what does a successful business model in the interactive broadband services market look like?*

Accenture answers this question by broadening the view of the individual market players beyond their traditional core business. From the business perspective we give advice for a successful take-off. The answer to this question guides us through this book. However, we start by explaining our understanding of the new industry structure as we see it emerging today. Based on this, we will go on to concentrate on revenue streams and business models.

We understand the interactive broadband service business to be part of the entire media and entertainment industry and not a pure telecommunications business. Therefore, the media business mechanics including the Get Audience/Sell Audience logic offer the appropriate framework for analysing issues and success factors in this attractive new market. This framework, introduced in chapter 2, is grouping all media business activities in the two main components Get Audience and Sell Audience. The Get Audience realm is to be understood as the challenge to acquire and retain customers, while the Sell Audience focuses on the necessity to further exploit the existing customers commercially. The framework is used as the basis for building up an understanding of this industry.

The interactive broadband evolution's impact on the Get Audience businesses is analysed in detail in chapter 3. Here, the media company's traditional core businesses are affected. Content production will develop a new and adequate form of storytelling in interactive broadband media - thus penetrating good revenues and profits. Content aggregators will need to understand customer intentions and habits in order to successfully target the right content to diverse interest groups. Only a few branded and trusted multi-channel access portals will survive. Content distribution is expected to become a commodity, with limited business opportunities for pure distributors relying on their original "bandwidth for money" business model.

Chapter 4 provides an overview of the development in the Sell Audience realm. Bearing in mind that Get Audience excellence is one prerequisite for additional Sell Audience revenues, three markets will be further discussed. So, a new quality of customer insight will be the enabler for all sorts of personalised services. Another revolution is set to take place in advertising, where today's interruptive ads will loose their dominating role in exchange for permission marketing and other innovative forms of targeted advertising. Among the key findings, synchronising the form of targeted merchandising and advertising with the offered content and the gathered specific customer interests will be a key success factor.

The understanding of this new industry structure opens up the view of possible existing and new revenue sources. Chapter 5 sums up all these identified revenue streams. Here, we give insights on how we expect future companies to actually earn money in this new business. The most promising revenue sources are presented. Intelligently combining market insights and revenue opportunities, we draw the conclusion to answer our guiding question. So, to end with, we will present implications and draw a design of possible business models that will be successful in this newly emerging environment. Also, we will include a scenario showing how the whole market is expected to develop.

Broadband media company profits will shift from the traditional "Get Audience" to the "Sell Audience" side of media business mechanics. Transactions will become the third pillar in the business.

2   "Get & Sell Audience" Basic Mechanics of All Media Business

With reference to our study "Reinventing Cable-TV Business" - in which we looked at the history, market and trends in CATV in August 2000 - the interactive broadband market potential seems to be huge: according to Datamonitor, interactive broadband services will grow into a $37 billion industry in 2003/4 in Western Europe, where at least one third is expected as *new* revenues.

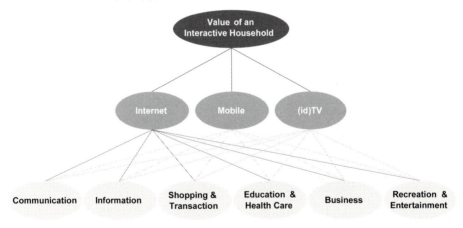

Chart 2: Value of an Interactive Household in Germany

The German interactive household has an estimated average annual value of approx. $1,250 in 2001/02. In the future, the interactive household is expected to spend more money via interactive digital devices, with its average value increasing to about $3,750 in 2005. To address this new market, content is needed in a way that fits the formats and business mechanics of the interactive broadband world.

Now, the question is how to exploit this potential. To answer this question a basic understanding of the interactive broadband media industry is necessary. The framework helping us to gain this understanding is the "Get & Sell Media Mechanics". This chapter introduces the framework and shortly discusses an array of issues along the "Get & Sell Audience" logic of broadband media business, currently under debate in media and telecommunications management circles.

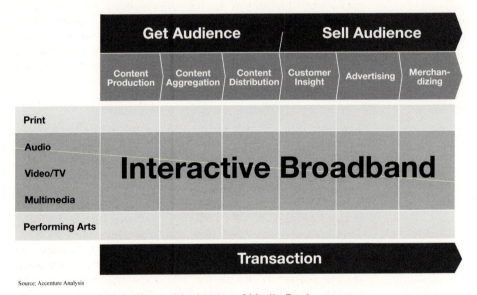

Chart 3: Get & Sell Audience Mechanics of Media Business

In our understanding, interactive broadband services are part of the media business. As illustrated in Chart 3, interactive broadband will merge the possibilities of audio, video and multimedia, offering two-way communication with fast access. Like in any other media business, therefore, broadband media revenues stem from two areas: "Get Audience" and "Sell Audience". Each consist of three parts - called markets. Content is the condensation kernel of the media business, it fascinates and "gets" the audience, thus forming the "Get Audience" - the content sales market. However, the markets of the "Get Audience" side are content production, content aggregation and content distribution. Yet the original content market profits are shrinking - mainly due to the effects of free Internet content. Thus, the importance of the "Sell Audience" markets grows dramatically. Without the revenues from customer insight, advertising and merchandising, only a few media companies could deliver profits. In the following the framework will be described in an overview.

In addition to the Get Audience and Sell Audience revenues mentioned, transaction is the third wide field of revenue opportunities within the interactive broadband environment. Transactions are present in all six markets of this logic. Therefore, an appropriate and combined discussion is incorporated into chapter 5 on revenue streams and future business models.

2.1 "Get Audience": Content Production, Aggregation and Distribution

The "Get Audience" contains three markets, as introduced in the mechanics of media business before: (1) content production, (2) content aggregation and (3) content distribution. Broadband content production will develop in pace with the technical & interaction opportunities in special interest areas like education, medical information, business matters, travel and shopping, as well as in the areas of general interest. The most promising revenue generator will be content production, followed by the content aggregation brand play. Content distribution will become a commodity.

Content Production

Every week a magazine needs to fascinate its buyers via a front page, to inspire them to buy it at a news stand or keep them from cancelling their subscription. A newspaper needs to do this every day via its headlines, content, image and credibility. A free television or radio broadcaster needs to attract young upscale audiences, e.g. via breaking news, movies, popular TV series and new formats. The same is true for Internet sites: the battle for eyeballs is extreme - users must be attracted by convincing, personalised and easy-to-use content that delivers rational value. Even theme parks produce content: Disney develops exciting experience worlds featuring its characters and thus attracts thousands of families to its locations every day.

Free TV and free content on the Internet have changed the audience's mindset. General interest content has acquired the connotation of being freely available. For a content producer, generating revenues has become a challenge, having only few opportunities: selling high valuable content to end customers, offering content to aggregators and exhibitors, placing advertising/merchandising into the content and/or trading their content property rights.

Content Aggregation

TV channels, for example, are content aggregators: they aggregate 24 hours of TV programs every day. Newspapers are content aggregators: every day an edition full of assorted and categorised news and opinions is published. The same is true for magazines. Internet portals are content aggregators: out of the streams of agency news and other information sources portals pick content relevant to their respective audience, elaborate and tailor it for interactive use and value delivery then present it to their audience as sorted content bundles. Interactive broadband portals serve the same end with the opportunity to do this on a different aggregation level: aggregating not only information but also different formats.

Content aggregators provide context. They bring structure and oversight into the overwhelming variety of news, information, transaction/interaction and other offerings. Advanced (interactive) content aggregators thereby adapt to the individual habits and preferences of the audience or allow manual (interactive) personalisation. In general, all content aggregators can be called portals.

By the way portals aggregate their content, they develop a tendency, a specific way of looking at news and information that stipulates and insinuates a certain point of view or value set. This tendency is a characteristic of media and can be used for differentiation and the development of trust in the audience. Trust in a portal is the most compelling reason for audience loyalty and a major element in the branding process. Branded Internet portals like AOL, T-Online, or Yahoo! have the opportunity to transfer their brand to the world of interactive broadband services.

Content Distribution

Newspapers need to be flung to the doorstep at 5.00 a.m. every morning or lie at the news stand at 6.30 a.m. to be easily grabbed by metro passengers on their way to work. Magazines need to be securely distributed every week to the homes and care has to be taken to determine the ideal numbers of copies to distribute to news stands and other points of sale in order to reduce lost sales on the one hand and returns on the other. In radio, television and Internet media, distribution or exhibition involves the operation of complex wired and wireless networks. Issues comprise the optimisation of technical facilities such as play-out centres, uplink and downlink stations, terrestrial, satellite, telephone and cable infrastructure and operations. For movies, the exhibition facilities are cinemas and home entertainment centres - increasingly enhanced with sophisticated new technologies, like surround sound and Imax screens, to enhance the media consumption experience. Several industries - e.g. brown goods, telecommunications infrastructure, wholesalers and retailers - make a big chunk of their revenues and margins with this part of the media business, which is usually not in the hands of the content producers or aggregators. For classic and new media, the distribution business has already become a commodity and prices tend to fall, especially in the area of digital narrowband content distribution.

Broadband content distribution - today still mainly the Internet, using technologies like cable, satellite or xDSL - is in its early stages and access premiums are therefore still paid. The supply of Pan-European capacity is expected to grow by 63% annually until 2005, when it will reach a total of about 13,000 Gbit/s. In the long run, broadband bandwidth will cease to be a limitation factor. The access possibilities are also developing rapidly. In Germany, the number of interactive digital TV (idTV) users is expected to grow by 38% annually over the next 4 years and the number of Internet users is still growing by 11% annually. In any case, content distribution is expected to become a commodity with price battles arising soon. Hence, in order to achieve a secure position in the market of interactive broadband content distribution, time-to-market is of ultimate importance - successful distributors use a first mover advantage to conquer and buy themselves into this developing European market. Slow movers will have to compete in niches or on price.

2.2 "Sell Audience": Customer Insight, Advertising and Merchandising

Once a medium - e.g. a magazine, newspaper, radio or television station, an Internet portal or a periodic event - has a stable audience and can be described in demographic and interest-related terms, possibly even down to the individual, the publisher is able to sell contacts to this audience for advertising and merchandising purposes. Facing decreasing "Get Audience" revenues, this side of the media business mechanics gains importance and profit impact. Therefore, integrated media companies must concentrate their attention on the development of ideal presuppositions for "Sell Audience" revenues, too. Three areas need to be developed: customer insight, advertising and merchandising.

Customer Insight

Customer insight is the centre of gravity on the "Sell Audience" side. Here, all information about the audience in general and individual customers is gathered systematically, leading to superior understanding of their demographics, habits, trends, lifestyles and attitudes. Each interaction with customers offers the chance to learn more about them. Customer insight development is an ongoing process of improving and updating audience profiles. The supporting means for customer insight are CRM (Customer Relationship Management) system tools.

Founded on their subscriber administration systems, most media companies have recently invested heavily in CRM technology.

Customer insight forms the basis for knowing the customers' needs, for segmenting the audience to enable companies to target the different segments with suitable advertising and to tailor merchandising offerings, as well as for identifying the value of individual customers. These basic achievements are extremely valuable because on the one hand they help to attract and retain large and stable audiences and on the other hand they are the basis for increasing Sell Audience revenues.

Advertising

Meeting the requirement of a stable and describable audience, value is created by selling contacts for this addressable audience to interested advertisers (and merchandisers). Originally advertisers bought space or time - separately or in media combinations. With improved technologies at hand, the media of today offer an array of services for all aspects of advertising, such as coupons, add-a-card, add-a-CD, personalisation to subscribers, selective printing and binding with high circulation or regionalised magazines and even Customer Relationship Services in cases where the audience responds with coupons or postcards from printed media, replies by e-mail or clicks through from Internet sites - all this more or less on the basis of interruptive advertising. Also product placement within the content area of the medium and sponsorships are playing an increasing role in selling audiences.

In the world of interactive broadband media, permission advertising will replace or at least supplement interruptive mass ads, be targeted along customer habits and tailored to interactive content. The advertising challenge will be to offer customers what they want when they want it.

Merchandising

In advertising, the medium offers its audience to advertisers for information about their products and services. Merchandising goes a step further. The medium offers and sells merchandise or services under its own name, co-branded with a producer, retailer or character, or in other innovative forms of cross-marketing. The revenue model is transaction-based and uses every possibility of transmitting brand, image or other intangible values to other products or services. The medium receives a fee for each sale. In the US, Martha Stewart has built a brand that stands for "Home and Living" and attracts millions of US Americans every day via its presence in newspapers, magazines, radio, television, the Internet, audio tapes and video tapes. Martha Stewart uses her brand and her media presence to recommend and sell products and services.

In the interactive broadband world, targeted merchandising will focus on merchandised products, recommendation and co-branding, along with the brands established around aggregators and content. Thus, merchandising will develop into a strong source of profits.

2.3   Revenue Shift from "Get" to "Sell Audience"

Every medium is inclined to skim maximum revenues from both the "Get" and the "Sell Audience" market in order to refinance investments and secure a satisfactory profit margin. In recent years, however, "Sell Audience" revenues have become more important - also in the broadband media. The introduction of the Internet to broad audiences and the availability of numerous free TV stations have made content sales to the end customer even more difficult. The Internet and private TV provide content free of charge for many areas of interest. Depending on culture and habits, this has influenced the general willingness of consumers to pay for content. For many media forms, the "Get Audience" becomes more and more a necessary presupposition to the "Sell Audience" without a significant revenue contribution of its own.

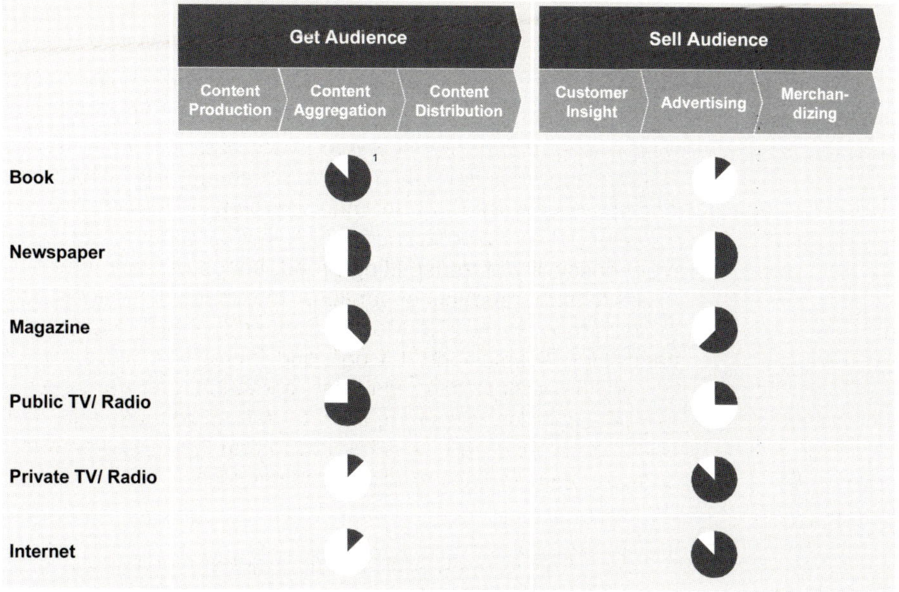

	Get Audience			Sell Audience		
	Content Production	Content Aggregation	Content Distribution	Customer Insight	Advertising	Merchan-dizing
Book						
Newspaper						
Magazine						
Public TV/ Radio						
Private TV/ Radio						
Internet						

Source: Accenture Analysis
[1] Filled Area Indicates Revenue Source

Chart 4: Financing Content by Different Types of Media

As shown in Chart 4, most media refinance their content through revenues from advertising, merchandising and customer insight (Sell Audience). This is the case for broadband media, too. Yet this is not a new phenomenon from the interactive broadband world, since it has been known for decades in printed media and the private sector of TV and radio broadcasting. While books and public TV are still mainly financed with content sales, young media such as the Internet and private TV generate most of their revenues from the Sell Audience side. And, beside being an excellent manager of these markets, everyone has to bear in mind that these Sell Audience markets will only pay if the basis – which means the product, service or brand – on the Get Audience side is at least in a good shape. This is the prerequisite for any Sell Audience revenues.

Also, in contrast to the described shift from "Get Audience" to "Sell Audience", this does not imply that it is impossible to run profitable businesses based on content sales. Especially the content producers are in a good position. Two factors must be considered. Firstly, content does not need to be paid by the end customer. Content producers selling their content to aggregators or distributors must not be affected by the end customers' limited willingness to pay. Secondly, this limited willingness to pay is no "law of nature", it is a question of value for money. When the customers see that the offered content is really adding value, they may well be willing to pay a fair and reasonable price. However, we will need to keep in mind that in most cases "Sell Audience" revenues will be even more important in the future than they are currently assessed, whereas "Get Audience" revenues will be reduced.

In addition to the Get and Sell Audience activities, there is a third powerful opportunity for additional revenues out on the market. Our analysis shows that by 2005, transactions involving shopping, paid information and communications will constitute the most requested services in terms of value. Because this affects all six markets, it will be discussed in section 5.

On the "Get Audience" side, content production and trading will remain a profitable business - content aggregation will become a brand play and content distribution a commodity.

3   Get Audience

Now, we start discussing the Get & Sell Audience framework in more detail beginning with the Get Audience side. Each of the three related markets will be discussed separately to understand the specific dynamics of this broadband media market.

In 2004 - we suppose - the sweet spots in the "Get Audience" part of idTV and broadband portals tend to be content production and content aggregation, leaving behind digital or physical content distribution, such as selling subscriptions or providing broadband access. To the contrary, companies that invest in technical broadband infrastructure might experience similar difficulties to ISPs (Internet service providers) did in recent years. In the future, each individual audience member will be able to choose a means of access to a broadband portal - idTV, Internet or mobile - from a number of options, considering situational and convenience factors.

A prediction shows that content producers will develop a new and adequate form of storytelling in interactive broadband media - thus penetrating good revenues and profit. Content aggregators will develop their portals towards multi-channel access and their business will become a brand play, while

17

content distribution is destined to become a commodity. The following three chapters illustrate the evolving broadband landscape in production, aggregation and distribution and introduce the most significant players.

Producers will develop a new and adequate form of storytelling in interactive broadband media to exploit the full market potential - thus penetrating good revenues and profits.

3.1 Content Production

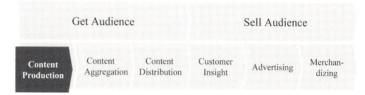

The first market on the Get Audience side is content production. This chapter introduces content production in interactive broadband media and highlights broadband storytelling, steps of technological enabling and consumers' willingness to pay. The main aim is not to discuss content production in general but to concentrate on the specific issues related to the production of interactive broadband content.

3.1.1 Interactive Broadband Content Production

The production of attractive content forms the basis for each and every media company. Attractive content is what is needed in order to attract the customers: Magazines need to fascinate readers with their front pages, Internet sites are faced with an extreme battle for eyeballs, and free television broadcasters try to attract their key audience via movies, popular TV series and new formats, to name just a few examples.

The art of storytelling is different for each media form - such as newspapers, magazines, books, audio books, television and Internet sites - and the same goes for interactive broadband content, too. This is why an exceptional newspaper journalist may be a poor television reporter and vice versa. Every form of storytelling has its own specific design and storytellers. There are key rational and emotional value expectations among the audience that make up the differences for successful storytelling in each media format. The art of storytelling itself is about addressing these value expectations and exhausting the specific media form to the fullest - it is not about copying content from one media form and pasting it into the other.

Attracting customers is essential for the new interactive broadband services, too. The difference is that broadband services cover a wider range of the communications spectrum than traditional media do. TV, on the one hand, as an example of a lean-back medium, is very much geared to human emotions, with pictures and sounds "going right to the heart". The Internet, on the other hand, is a very informative, rational, lean-forward medium, where people request specific information and emotions do not play a leading role. The interactive broadband spectrum, as illustrated in Chart 5, has the advantage of covering and addressing emotions and information at the same time, by combining the Internet and TV capabilities in one service. The customer gets the chance to choose between a very emotional application, a rational application or a combination of the two. Particularly good examples of this are e-learning and tutoring, described later in this chapter.

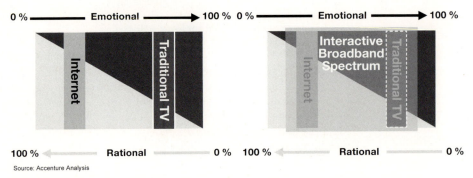

Source: Accenture Analysis

Chart 5: Emotional and Rational Content Develops Towards Multisensual Content

As mentioned before, the broadband communication spectrum will be larger than the spectrum of the traditional media. Beside this, the new broadband technology will enable full interactivity. Now, to exploit all these new possibilities, a new kind of

20

storytelling has to be developed. This kind will be directed to more than one sense, combining the traditional techniques with the new broadband possibilities. As a consequence, the content offerings will be adjusted and changed. Looking at Chart 6, it becomes obvious that the art of storytelling has to change from a more or less purely linear/rational Internet content and linear/emotional TV content to an interactive, rational and emotional broadband content - enabling the end users to literally interact with their broadband access device (should they so choose) - and therewith directing and creating their own program.

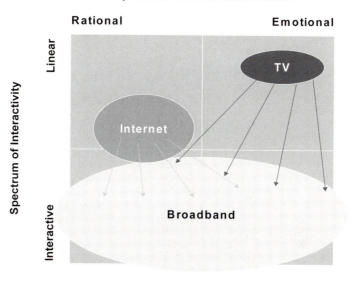

Chart 6: Broadband Widens the Range and Mixture of Traditional Storytelling Techniques

3.1.2 The Art of Interactive Broadband Storytelling

The fact that broadband content is able to use the full spectrum of communication and interactivity gives rise to a new art of interactive broadband storytelling. For all types of media, fiction and incidents these form the two sources for the creation or occurrence of content. As can be seen in Chart 7, fiction (e.g. a song, a book, an idea, a story, a vision, etc.) creates content by being launched and published. The other source is in form of an incident, which could be breaking news, weather, etc. Also, every piece of interactive content produced has its origin in one or the other source. The entire content as a whole - regardless of whether the content is traditional or broadband - can be split into six areas demanded by the customer.

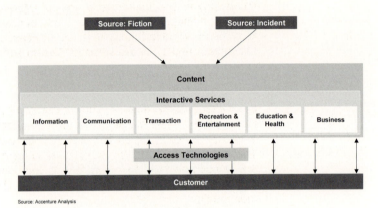

Source: Accenture Analysis

Chart 7: Sources and Clusters of Content

The preparation and development from the pure piece of information to broadcast content involves various different steps, which depend on the medium on which the content will be broadcast. For example, TV content requires motion pictures to be prepared, Internet requires programming in HTML and magazines require articles to be written. No matter which access medium is used, traditional or broadband, there are different "factors" that make content fascinating:

- The "art" of storytelling

Each medium has its own tricks and forces. To exploit the full potential of each medium, the art of storytelling is the craftsmanship and knack of presenting the content adequately. Each form of content requires another form of storytelling. Due to the fact that content production is expensive and time-intensive, the same content is reused in various media, whereby the art of storytelling is always adjusted to the medium. As technology is enhanced, the degree of interactivity of the content increases. Requirements for successful interactive broadband will also be technical capabilities, a critical mass of consumers and business opportunities to sell or refinance the content

- The storyteller

Storytellers can be authors, show masters, etc. - depending on the medium. Their challenge is to "transport" the content as efficiently as possible on the kind of medium they are using. They have to fascinate the users - to bind them to their program and encourage them to "subscribe" to their show. Ideally, they have to exhaust all the possibilities of storytelling offered by their medium.

- Pictures

Choosing the right kind of pictures for the different stories and the different media is an art in itself. Pictures play a key role in different media - not only in TV and in the movies, where the dependence on pictures is obvious. Picking out the pictures, which are put beneath an article in a newspaper, magazine and website, is also extremely important, because they will stay as an assimilation in people's memories for a long time.

- Sounds

Music and sound are very emotional factors that dispose of a very high recognition frequency. Nothing stays in a human memory longer than music and smells. Good music is an incentive for people to enter a store, listen to advertising or watch a movie.

- Graphics

"A good graphic is better than a page of explanation" is a saying that is very applicable in this context. Graphics enable the reader or end user to visualise very complex and difficult ideas and thoughts, by using colours and trends so that data becomes clearer and easier to compare and more attractive.

- Colours

Good composed colour arrangements are eye catchers. People have a high affinity for colours – colours transport emotions. Therefore magazines, Internet pages, etc. tend to stick with the same colour range so that people recognise a brand name immediately and start to feel "at home" when reaching a site or opening a magazine.

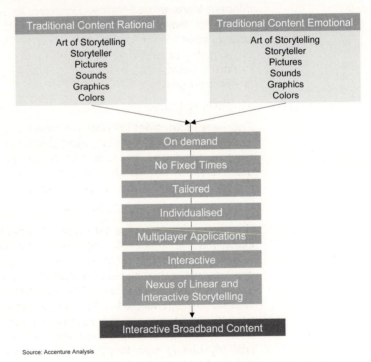

Source: Accenture Analysis

Chart 8: Evolution from Traditional to Broadband Content

Keeping in mind these six factors essential for all kinds of media content production - regardless of whether traditional or broadband - the next step is to find specific factors in which broadband content differs from traditional content. There, the key question is how to attract and retain end users when designing broadband content. To a certain extent, interactive content is the development of traditional content, enabled through the existence of a feedback channel and a fast broadband connection with a minimum of 1.5 Mbits/s downstream. This opens up new possibilities and developments, as illustrated in Chart 8. Important features of broadband content will be:

- On demand

End users will be able to choose their desired content on demand. That implies that they can choose the time when the different kinds of content are delivered. In addition, they can determine their own level of involvement, from lean-back entertainment - enjoying the multimedia capabilities of broadband - to being very involved in highly interactive lean-forward activities such as e-learning, quizzes, game shows and multi-player games.

- No fixed times

End users can order content whenever they choose without having to stick to program schedules. Due to the feedback channel, users can rewind a program as often as they like in order to watch or "use" the content over and over again - very similar to what we know from our VCRs nowadays.

- Individualised/tailored offerings

Due to the EPG and the narrowcast functions, content will be individualised. Content offerings will only be proposed to special target groups and people of interest. The individualisation creates a win-win situation between the content provider, aggregator and the end user. The end users receive targeted content offerings tailored to their particular interests, while the content aggregators receive valuable customer data. They can use this data for their own planning, as well as turning it into cash by selling data to advertisers and merchandisers (see section 4.1.3).

- Multi-player applications

Today, various players can connect their applications, such as Play Station and Game Boy, via cable to play with each other. In future, broadband will allow interaction between various players from all parts of the world, independently from their access device.

- Interactive

Broadband content will be interactive. Through the feedback channel the end user will literally be able to interact (i.e. to ask questions, order special content or products and participate in live acts).

- Nexus of linear and interactive storytelling

As described, broadband content enables you to combine linear and interactive content, which makes broadband content and applications so extraordinary interesting. Broadband content captures more senses, so that the right-hand side (emotional centre) as well as the left-hand side (rational centre) of the human brain are participating in the process of perception.

3.1.3 Content Production Develops Along Technical Network Development Stages

The full range of interactive broadband content can only be developed in a meaningful way in line with the development of the technology platforms delivering and providing this content. Only when the diverse access channels have been fully upgraded to the necessary broad frequency range and a broadband back-channel becomes available will the content offerings migrate from a "push"/broadcast mode into an interactive "pull" or on-demand mode. The most probable development path of idTV, broadband technology and its content over time can be sketched roughly by describing four stages of technical development, shown in Chart 9. The dates indicated are valid for Germany, Austria and Switzerland. This means that content development comes hand in hand with technical network development.

Platform 1	Platform 2	Platform 3	Platform 4
Digital TV Broadcast	**Enabled dTV**	**True idTV**	**Open Broadband Standard**
Broadcasting	**Broadcasting Plus Some Interactivity**	**Broadcast Plus Narrowcast Content On Demand**	**Broadcast and Narrowcast Hemispheres Fully Developed**
• Launch of Firstidtv & Broadband Entertainment Platform • Free TV/ Pay TV Program Bouquets • Basic Push Services • SMS/ Ticker Services • Cross-media Links, URL Links • Field Test Kabel Berlin • Special Interest Channels • Opening of Currently Pre-dominantDbox for MHP Besides Embedded Betacrypt	• Establishment ofidTV/ Broadband Platform Brands in Germany and Europe • Program Bouquets • Time-shifted TV & Personal VCR • Real-time Feedback • Voicemail/ E-mail, Chat • Fast Internet Access • Choice of Contents (Movie Archives, ZDF, ARD)	• Branded idTV/ Broadband Platforms with Personalization Offerings • Personal Interactive Services • Time-shifted TV & Personal VCR • True VoD, Video Download • Personal Push & Pull Info Services • Video Conference • MHP xDSL Standard	• Multi-channel Access & PersonalizedidTV/ Broadband Platforms With Automatic End-user Device Detection (TV set, PC, Handheld PDA, WAP/ UMTS phone, etc. with Technical Functionality as a Commodity) • Global Content Retrieval • Open for All Content Providers • Brand and Marketing Most Important
Existing Set-top Boxes	**Boxes With MHP Standard**	**Fully Interactive MHP-Based Boxes**	**Fully Interactive MHP-Based Boxes**
2001	**2002**	**2003**	**2004**

Source: Accenture Analysis

Chart 9: Development Stages of Interactive Broadband Content

- **Stage 1 in 2001 enabling broadcasting and streaming:** In front-end TV, digital TV remains a bandwidth-saving transfer technology for sequential TV programs. Network and set-top boxes do not allow major interaction between the customer and the content or the service provider via the TV set. All TV offerings will be broadcast-based. Basic push services to customers, cross-media links (SMS, ticker services), free TV and usual pay TV are possible. In the PC front end, broadband content will still be hindered by a lack of bandwidth. It will be possible to develop small-size video with interactive features but not to convey "emotional" value or experiences. Business applications are lagging behind, before a critical mass of customers is reached to drive large-scale applications.

- **Stage 2 in 2002 enabling dTV:** In front-end TV, network and set-top boxes will begin to be sufficiently enabled to allow some interactivity between the audience and content/service provider; e.g. personal EPG (Electronic Program Guide) with a personalised entry screen, personal VCR services, time-shifted viewing, hosting/retrieving of third party (retailer/media) content and a few true on-demand offerings like down-streaming of selected movies from archives into the set-top box, voice & e-mail, chat, broadband Internet access, meta-information with video, individual feedback on live TV programs. In the PC front end, initial, fully interactive broadband functionalities are introduced.

- **Stage 3 in 2003 enabling true idTV:** In front-end TV, initial fully interactive idTV offerings will be available. Network and set-top boxes are sufficiently equipped with server infrastructure and functionality for additional services, like time-shifted free TV and pay TV (broadcast), full real-time on-demand services, personal push-and-pull information services and video chat. In the front-end PC, fully developed interactive broadband contents & services in both hemispheres are available. Fully interactive dTV will allow personal interactive services and be the true starting point for branded idTV platforms with personalised and tailored offerings. To achieve the critical mass in a consumer technology, 15 to 20% of the household population need to request the service. Most European countries are expected to reach the critical mass for digital TV before 2003.

- **Stage 4 in 2004 with Open Multi-Channel Access Broadband Portal:** At this time the difference between mobile, idTV-based and IP-based broadband portals will disappear. The mobile and Internet PC-targeted portal ap-proach and TV-centric approach converge to form one portal platform with joint branding and marketing. This broadband platform will be open for all on-demand services via diverse access channels. The mobile, TV and PC front ends can display the same content and service offerings. Brand, content, cost effectiveness and handling convenience are key success factors. Technical access and functionality are commodities and bandwidth will no longer be a limiting factor. In 2004 we will face the open broadband standard with the possibility for true multi-channel access. Platforms with this standard will allow multi-channel access with personalised user identification, independently from the device used. This open standard will foster competition among brands and allow open access for all content providers.

3.1.4   Different Areas of Interactive Broadband Content

All kinds of interactive broadband content can be organised into different groups:

- **Communication** is a basic service requirement that must be satisfied. The ability to receive and send voice or data messages to any location has led to new means of communication. Newsgroups, chat rooms, e-mail, voice-mail, buddy nets and instant/unified messaging services, video-conferencing, video chat and IP telephony are becoming components of everyone's communication needs.

- **Information:** Personal data & archives. New interactive services that address standard everyday issues and the behaviour of the average household can boost demand and, in addition to the individualisation feature, increase user "loyalty" to broadband/idTV content. Examples include: interactive Juke Box, My Photo & Video Album, personal video files & production software (scripting own videos from the provider's content archive), personal program filing/VCR, as well as the traditional homepages and information sites. This section can be extended with services like tracking, finding, guiding and emergency.

- **Shopping & transaction:** Most examples of value-added services are the same as on the Internet. Home banking, home shopping and access to personal data & archives are services that deliver the ease of 7*24h accessibility tailored to the lifestyles of customers. But there will be a major difference in the richness of the form of storytelling. Transactions belong to the value-added services and include the ability to perform activities such as online shopping, home banking and paying bills. But factors like costs, range of goods available and the ability to compare prices among others are also influencing this content need. Typical examples are home shopping and home banking, auctions and classical e-commerce applications.

- Education & health:

  1) The demand for *education and tutoring* involves accessing learning materials, obtaining teaching information, participating in tutoring sessions or being able to perform research in a simple, comprehensible way and within an appropriate duration at any time. This can be achieved by a corresponding educational or academic channel or sub-portal, which combines the requirements of educational content and customer-driven handling. In Europe this is a new area that is in the process of being launched under the buzzword 'e-learning'. Several governmental and private initiatives promote the use of multimedia and technology at school and at home. In the area of education (e.g. a channel with offerings ranging from quick homework help, repetition and tutoring to comprehensive self-paced courses for all major disciplines at school, university and business), access to learning materials, obtaining teaching information, live online participation in educational events or performing research can be explored. A starting point for such service could be the re-use of the rich educational contents of Telekolleg, a program that has been produced by the German ARD for decades. This content could be indexed, thematically portioned and sent on demand.

  2) *Health and Wellness:* The same scheme applies in health care: On-demand content e.g. about the most relevant illnesses and diseases and their treatment, courses/information on special health and wellness issues as well as actual features of professions in health care with information, communication and transaction offerings for health & wellness are bundled in a corresponding portal.

- **Business:** Personalised business information & news. The demand for business content is another upcoming star in the digital interactive world. It contains everything necessary for all kinds of small and home offices starting with tools for office management and ending with business software. Hence, the next major wave will bring business software, such as ERP (enterprise resource planning) software, into the content space on the network. The rise of

ASPs (application service providers) is one step in this direction. Personalised business information is the offering of compilations of profession-specific audio & video news feeds, which can be clicked from the portal shelf.

- **Recreation & entertainment:**

  1) *Games & Gambling:* Recreation and interactive entertainment demand constitutes a business opportunity for games & gambling leisure services, offering this service on demand with the ability to play with other players no matter where they are and which device they use. Referring to Datamonitor research, this market will increase to $7 billion by 2003. As we know from the use of personal computers, educational and recreational applications are key drivers for influencing the acceptance of technology. Broadband mobile, idTV and Internet portals are an ideal platform for a new breed of interactive games. As an example, family A in Berlin plays a football game against family B in Munich via the broadband portal platform - every family member takes over a role or player in his/her team. Electronic Arts, the world's leading electronic games producer, is in the process of developing a new generation of games especially tailored for use via Internet (in co-operation with AOL). No partnership with a broadband/idTV aggregator has been established. In the mobile realm, for example, NTT DoCoMo (iMode) has teamed up with Voicestream, AT&T Wireless, Télécom Italia, Hutchison and Sony (PlayStation).

  2) *Video & Audio-on-demand Archives:* This is the area for pure on-demand general and special-interest offerings, such as movies, audio books & plays, music down-loads, general & personalised business information & news. Music on demand will be used instead of buying a CD. Music will be downloaded in the form of MP3 files to be stored on the PC or MP3 player like audio books, or will be streamed down for immediate consumption without filing. This channel also provides fast digital access to thematic video archives of rights owners or distributors, like Kirch Media, RTL Group, etc. The advantages for the audience are fast access, own timing and convenience of purchase. On-demand video and video streaming will sustain triple-digit growth, from

250,000 units shipped by the end of 2000 to 8 million units in 2003, with key buyers expected to be hotels, large apartment buildings and even cruise ships.

3.1.5   **Requirements for Attractive Future Broadband Content**

As shown in Chart 10, successful interactive content meets the following criteria:

- *Cost efficiency*: Has to offer a good value.
- *User friendliness*: Has to be easy to use, understand and handle.
- *Advisability*: Content offerings have to be useful and have to make sense.

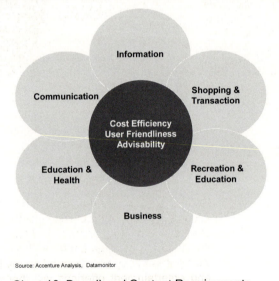

Source: Accenture Analysis, Datamonitor

Chart 10: Broadband Content Requirements

Comparing the content demand (measured in interactive time spent during a week online) of an interactive household in 2004 with its willingness to pay (measured in % of online dollar spending), as shown in Chart 11, makes it obvious that consumers are not willing to pay for services in the way they are demanding. For example, information is believed to represent 1% of dollar spending but 15% of total time spent online. Willingness to pay is driven by communication (39%) as a human basic requirement, representing 25% of the time spent online during a week. In the case of shopping (36%), it is obvious that little time is spent online but large amounts of cash are transferred during shopping transactions.

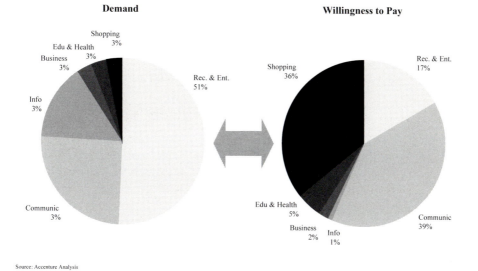

Source: Accenture Analysis

Chart 11: Demand vs. Willingness to Pay

Recreation and entertainment represent only 17% of the total spending of the interactive household, but, at 51%, VoD and games will be in demand in 2005. Education & health and business will be segmented applications (few hard users) with willingness to pay and demand nearly in equilibrium. Recreation and entertainment offerings will therefore rely heavily on refinancing by the Sell Audience.

3.1.6   Willingness to Pay

In order to be able to predict how much content of which kind is going to be requested, it is important to look at the 'need to have / nice to have' media logic shown in Chart 12 before being able to make any assumptions.

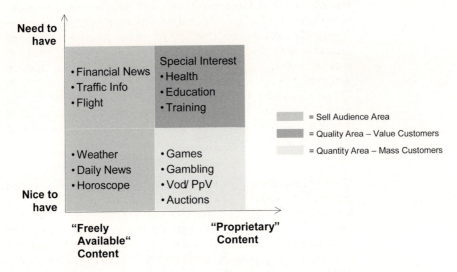

Source:AccentureAnalysis

Chart 12: Media Logic: Need to Have - Nice to Have

Broadband content can be classified in different groups. There is one group of content that is essential for its users to have - e.g. special books, applications for students, doctors and other groups of special interest. In the following model we refer to this content as "need to have". On the other hand there are content offerings that are "nice to have", like games, gambling and weather information, - things people request but they do not necessarily need. Those two groups include content that end users are willing to pay for and content they expect to get for free.

As a result of this model, there are different clusters which have to be approached from different viewpoints and in different ways: The left-hand side of the model is important when looking at the number of users in general, while the right hand side is interesting for companies offering broadband services, as they get paid for the content they are providing directly.

Having analysed the two different groups, we see two major opportunities: "quality and quantity". Quality is in the upper right-hand side. Information that is necessary for special interest groups - such as students of different academic fields (e.g. medicine, law, architecture) - who are searching for specific information they necessarily need and are willing to pay for it. These types of broadband services are seldom requested and only by persons with a special interest but at the same time promise relatively high turnover due to the willingness of end customers to pay.

Quantity: the second group on the lower right-hand side is also prospering, again due to the customers' willingness to pay for the services they are using. The difference to the categories presented before is that the content being requested does not fit the category "need to have" - it is only "nice to have". Thus, the customer's willingness to pay is a lot lower than in the quality category described above. Another example of the non-profitability of a service in this area is the streaming business. Customers use the service too little and are not willing to pay for this kind of service. The content offered in this area needs to be refinanced by the "Sell Audience" i.e. sold to aggregators distributing the content for free, using, for example, advertising and merchandising, to refinance themselves. A vivid example of this model is the phenomenon that end users access stock quotes, news and horoscopes on a regular basis, though they are not willing to pay for this.

Napster & the Need for Digital Rights Management (DRM)

The easy way in which digital content can be downloaded and distributed has demonstrated that the rights of off-line providers are fundamentally endangered in the Internet realm. The recent controversy within the music recording industry, dealing with the file-sharing community Napster, serves as a prominent example of the expanding phenomenon of online piracy. As a peer-to-peer network, Napster runs servers that direct more than 60 million inquiring consumers to the computers of other Napster users, from whom they can download songs they are looking for (in the MP3 format) for free.

By doing this, the Napster network provides the infrastructure for unauthorised use and reproduction of copyrighted music content, and causes harm to the music industry. As a result, rights holders lose substantial revenues due to illegal reproduction and distribution of protected content. While the music industry has taken legal action against Napster, Bertelsmann is teaming up with Napster to develop a means of making use of Digital Rights Management (DRM) as a basis for an alternative business model including fees for downloading music files.

The need for proactive approaches to online rights protection - collectively known as DRM - is obvious. The design and implementation of an effective rights protection solution is a challenging task, since DRM solutions have to meet very different needs: business needs, i.e. preventing unauthorised distribution and consumer needs, i.e. user-friendly interfaces.

Online rights protection begins before the content reaches the consumer. DRM Technology features at production level (e.g. watermarks and encryption) - though more expensive to implement - ensure more effective control over content than activities at the consumer side (e.g. copy control or password registration). On the other hand, control features at the consumer end (e.g. permission sets) used as market research tools offer customer data mining opportunities.

To serve specific safety needs, DRM solutions can include a combination of rights protection features. Additionally, a DRM solution can be used as a basis for alternative business models for online content providers, since market and consumer data can be turned into incremental revenues (e.g. by generating customer data, targeted advertising or market research). In a nutshell, the possibilities of DRM solutions extend far beyond mere piracy protection.

Main Content Production Revenue Streams

Sales of produced content to content aggregators, content distributors, and end customers. These revenues consist of both, monthly subscription fees and individual payments for VoD or PPV.

Licensing includes all revenues related to the sale of rights to other content producers and aggregators, especially in foreign countries. An example is Burda editorial, which sold the format and the rights on "Focus" to a Brazilian editorial producing and publishing the same magazine in Brazil today.

Syndication is the sale of produced and copyrighted content formats (e.g. Big Brother), movies, texts, or photographs to other, national or international editorials or aggregators.

Revenues by transactions as present in all markets, explained in detail in chapter 5.1.3.

> Content aggregators will need to understand customer intentions and habits in order to successfully target the right content to diverse interest groups - in the long run only a few branded and trusted multi-channel access portals will survive.

3.2 Content Aggregation

The second market on the Get Audience side is content aggregation. In this chapter we will discuss in more detail the situation of content aggregation in interactive broadband media. We will concentrate on the different development stages and the evolution of multi-channel access portals. Because in the content aggregation market driving traffic to the desired content and branding of the respective portal are seen as the two major key factors of success, the discussion will begin with these issues.

3.2.1 Why Do We Need Content Aggregation?

Content aggregation has played a significant role in the media business, since mass production of newspapers, magazines, etc. was introduced, because the aggregator is the "appearance" of every kind of media. The variety of different kinds of content asks for aggregation to deliver structured content and navigation help. The art of structuring content and offering it in a special way through storytelling to the customers gives rise to an aggregator's long-term goal to create a brand name that people identify themselves with. As an example, newspapers such as "The Financial Times" or "Frankfurter Allgemeine Zeitung" are bought because customers know that the newspaper will satisfy their need for high-quality, up-to-date information.

Successful content aggregators present content to consumers structured in a way that the consumers appreciate. This is especially important in the recent Internet age, where the variety of content is huge and the whole net appears like a jungle. The aggregator will own the customer's relationship by recognising the customers when they log on and tailoring content offerings to customer's habits, needs and preferences. In the future, multi-channel aggregators will provide access to the desired content no matter how the consumer accesses the aggregator's portal. The aggregator provides look & feel, individualises the content to the consumer preferences, provides functionality, presents the content in an individualised form which meets style preferences, gives a pre-choice of topics and also a tendency or point of view to the content in a comfortable way. This mixture lays the foundation for consumer trust in the portal. The consumers place their confidence in the aggregator - thus the aggregator is on the way to becoming a brand.

As a result, aggregators are brand names that stand for a special kind of storytelling, a special kind of opinion and colour and a special kind of structure. Everyone searches for and uses the aggregator they feel most comfortable with and rely on. Especially in the imminent battle between the various broadband service aggregators, branding will become a major success factor. In order to create a brand name and image, an aggregator has to offer content, presented in an orderly way so that end users know that the content and information they are looking for is offered and that they can navigate through the offerings, always knowing where to find what. Yahoo is one good example of a successful aggregator in the modern Internet business. Another is focus.de, the aggregator site of one of Germany's traditional media companies.

Also, the reason why aggregators are essential in the media business is related to the economies of scale that can be achieved by reusing and syndicating data. Traditional publishing companies, for instance, can use the researched and developed content not only for publishing it in the traditional media but also to offer it on their websites, enriching the content with more applications, more pictures, background, etc. As a different example, Yahoo has created a platform on which every kind of content is offered. Due to the diverse offerings and the high number of different websites and aggregators, the creation of a well established brand name during the first hours of broadband service is essential, since time-to-market is a key success factor - only the fastest will have a chance and only a few well-known names will be able to survive in the market, positioning themselves as an elaborated and advanced service to a large group of customers.

Development Stages of Content Aggregation

Broadband technology opens up new possibilities for content aggregation. Along with the technology development, different development stages of content aggregation can be observed, as illustrated in Chart 13. During the first stage, content gets networked. This means that, because more and more content is offered on the Internet for people to download, this content has to be clustered and structured, and a navigation help needs to be developed. Examples include: search functions and content directories at the very beginning of the stage and service and local content as the last elements of the "content gets networked" stage. The second phase of content aggregation is described as "Content gets personalised". In this stage end users are giving away their personal data in order to receive content designed specially for their specific interests. The idea is to end up with portals that are more or less individualised and personalised. Content tends to be more interactive and dynamic. The degree of customer relationship is increasing.

Whereas in the two stages just described above separate portals (PC, TV, mobile) still exist in parallel, the third stage, "content gets everywhere", recognises the user no matter which means of access is chosen. Aggregators will prepare access to the content tailored to customer needs and habits and specially prepared for the access device chosen. As a result, the content is totally personalised and accessible from basically everywhere and there is a high degree of customer relationship. Personalised multi-channel access portals are born.

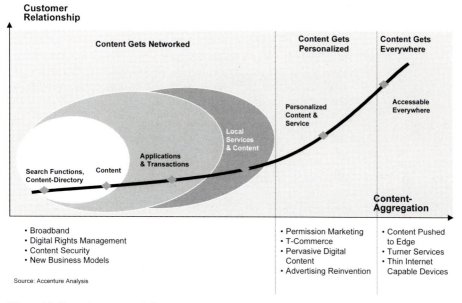

Customer Relationship

Content Gets Networked

Content Gets Personalized

Content Gets Everywhere

Accessable Everywhere

Personalized Content & Service

Local Services & Content

Applications & Transactions

Search Functions, Content-Directory

Content

Content-Aggregation

• Broadband	• Permission Marketing	• Content Pushed
• Digital Rights Management	• T-Commerce	to Edge
• Content Security	• Pervasive Digital	• Turner Services
• New Business Models	Content	• Thin Internet
	• Advertising Reinvention	Capable Devices

Source: Accenture Analysis

Chart 13: Development of Content Aggregation and Customer Relations

In a few years, when we turn on the idTV in the morning, we will see the news the way we configured our aggregator. We will get offered short clips of our favourite sports news, see political and economic news on broadcast and have our portfolio in another window. While driving to work we can access our aggregator with a mobile device (e.g. PDA or mobile phone) and search and download the latest hits. When we turn on our PC on our desk at work, we will enter the virtual world again using the same aggregator. No matter where we are, in a hotel, train, at home, even in a plane, no matter which device we use, a PC, PDA, mobile phone or TV, we will have access to customised and tailored content offerings This situation is illustrated in Chart 14.

Content Gets Networked & Personalized

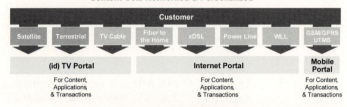

(id) TV Portal	Internet Portal	Mobile Portal
For Content, Applications & Transactions	For Content, Applications & Transactions	For Content, Applications & Transactions

Content Gets Everywhere

Multi-Channel Access Aggregator

Source: Accenture Analysis

For Content, Applications & Transactions

Chart 14: Content and Access Development

3.2.3 Evolution to MCAPS and Relevant Examples

Consumers want to access their favourite aggregator with customised offerings no matter where they are and which device they use. Audience members will want to individualise the desired content to cover their specific needs and to reduce complexity. The most popular content will be book-marked for quicker retrieval. Although there is no true Multi-Channel access portal (MCAPs) yet, initial players are already evolving. They will gain in importance for the audience as the number of digital broadband channels and offerings increase. They are also important for pay-per-view, video-on-demand transactions and other interactive services like electronic business transactions. In this context, layout and handling of EPGs will become important success factors - layout, functionalities, offerings, recommendations and ease of handling create an experience for the audience and a brand. A comprehensive and individually adaptable EPG is a key factor for success, as is the early establishment and promotion of this brand, ideally by allying key players of the three different market sectors to perform as one aggregator. In order to be prepared, many companies are beginning to establish alliances and networks in every access area. NTT DoCoMo and Vodafone/Vivendi, in particular, are seen as first movers. The technology for "true" multi-channel access will not be available until 2004, but the meaningful players are already forming alliances and partnerships to offer and aggregate content via the Internet, (id)TV and mobile.

Examples of this situation (with no claim to completeness) are depicted in Chart 15. Bertelsmann - with CLT-UFA as a large content producer - and the RTL Group as an (id)TV aggregator is teaming up with RTL World, Terra Lycos and Napster in the Internet realm, as well as Telefonica, 3G and Sonera in the mobile realm.

AOL Time Warner has very international co-operations in the mobile sector with NTT DoCoMo (iMode), AT&T Wireless, BT, Telecom Italia (which has a 25% share in Mobilkom Austria A1), Sony Playstation and Hutchison. On the (id)TV side AOL Time Warner is partnering with CNN News Group. AOL Time Warner also contains two content producers: Warner Brothers and Warner Music Group.

Deutsche Telekom, which recently acquired Voicestream, is building an "in-house" solution. In the (id)TV realm, Deutsche Telekom is partnering with ARD digital, ZDF Vision and Premiere World, in Internet T-Online, Bild.t-online.de, and in Mobile T-Mobil D1, One2One (UK) and MaxMobil (Austria).

Kirch Media - a large owner of rights and content producers like Janus and Taurus - have opportunities for teaming their (id)TV activities at ProSiebenSat.1, Home Shopping Europe, tm3, Premiere World, N24 and DSF with their Internet activities Sat1-Online, ProSieben-Online and RedSeven.

Vivendi and Vodaphone D2 are building a multi-channel access portal called Vizzavi (starting with the Internet), partnering with Swisscom, debitel, Airtel, iusacell, Ericell and Japanese Telecom in the mobile field and with Vivendi Universal and Canal+ in the (id)TV aggregator field, having Universal as an integrated content producer.

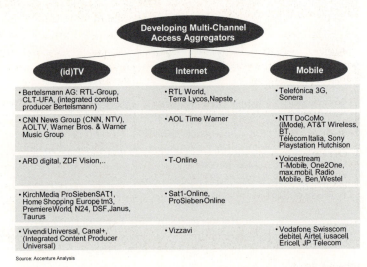

Source: Accenture Analysis

Chart 15: Examples of Networks Developing around Multi-Channel Access Aggregators (with no Claim to Completeness)

3.2.4 **MCAP's Offering in a Broad- and Narrowcast Hemisphere**

Broadband content aggregator's offerings will develop towards a channel structure with a broadcast (one-to-many) and a narrowcast (one-to-one) hemisphere - both with interactive features. This situation is illustrated in Chart 16.

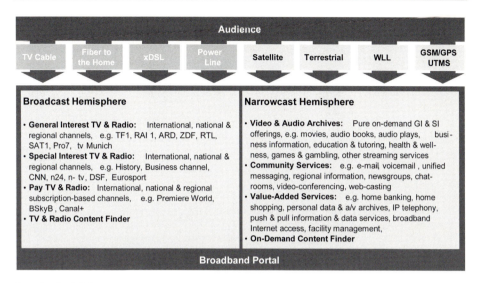

Source: Accenture Analysis

Chart 16: Broadcast and Narrowcast Hemispheres

The broadcast hemisphere includes linear TV contents (e.g. general interest TV, special interest TV, pay TV and TV content finders) plus basic convenience features, like time-shifted viewing and targeted advertising.

The narrowcast hemisphere includes personalised interactive on-demand contents and services for individual users for use in their own time and at their own pace. This offers a platform for targeted interactive advertising and personalised permission marketing. Video & audio archives, community services, value-added services and on-demand content finders will also be part of the narrowcast hemisphere shown in Chart 17.

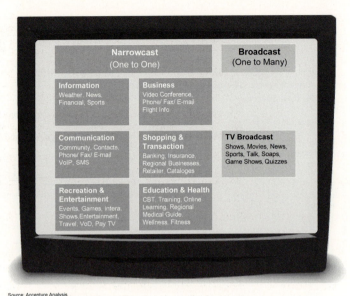

Source: Accenture Analysis

Chart 17: Interactive Broadband Services Bouquet

3.2.5

MCAP's Struggle and Partnering for a Predominant Brand

In order to be successful and create a stable positioning within the new broadband market, MCAP's have to consider various factors when designing and offering their services, as illustrated in Chart 18. First of all, aggregators need to establish a strong brand. In a growing market it is important to create trust and gain the stake. Moreover, the service has to be scalable, personalisable and has to offer a look and feel that satisfies the customers' preferences and interests. Customers have to be triggered with incentives and enticed by rewards in order to bind them to the MCAP brand and win them as long-term clients. Essential for the long-term success of the aggregator is a sound business model, the ability to really create customer loyalty and satisfaction by offering a strong content with killer applications.

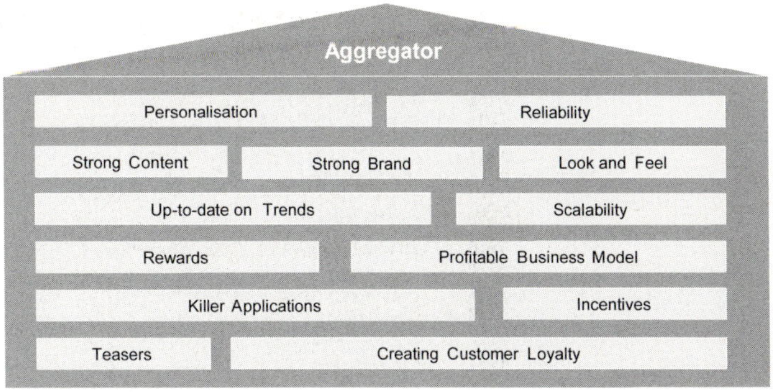

```
                        Aggregator

    Personalisation              Reliability

 Strong Content      Strong Brand      Look and Feel

    Up-to-date on Trends              Scalability

    Rewards              Profitable Business Model

    Killer Applications              Incentives

 Teasers         Creating Customer Loyalty
```

Source: Accenture Analysis

Chart 18: Successful Content Aggregation Building Blocks

As described earlier, partnering will be one of the milestones to success when creating a successful aggregator. Combining three leading companies of the "access areas" PC, TV and mobile to create a joint service as an MCAP will mean a major advantage in a business where time-to-market and branding are the key success factors. Partnering therefore becomes necessary.

3.2.6 Aggregators Can Act as Market Makers or as Entrepreneurs

Looking at the relationship between content aggregators and content producers, aggregators can act in two different ways. On the one hand, they can act as market makers. In this model, content producers provide their content to the aggregators, who "make the market" for the content producers by enabling a broad audience to use the provided content. They usually generate their revenues on a revenue-sharing basis with the producers. A prerequisite for this revenue-sharing is that the end customers pay for the content individually (either pay-per-view or on demand).

On the other hand, the content aggregators can act as entrepreneurs on their own responsibility. Following this approach, aggregators buy content from the producers and pay the corresponding license fees. They then distribute the content to the end customers. Here, they run the risk of having paid a large amount of money for content that is not requested as much as expected, while they have the opportunity to achieve higher profits if the content is sold better than expected. In addition, content aggregators are completely free to decide how to put the content on the market. It is not necessary for the end customer to pay for the content individually. To the contrary, the aggregators may provide the content for free, generating their revenues in other markets based on the Sell Audience or transaction activities only. These activities will be discussed later in this study.

NTT DoCoMo - Pioneer of Mobile Multimedia Portals

iMode - Mobile Internet Access and Service

iMode is a mobile Internet access and service system owned by NTT DoCoMo. Users access the Internet directly from an iMode-compatible mobile phone e.g. with the possibility of transferring pictures and short video clips of good quality. In addition to voice transfer and mobile Internet access, iMode-enabled handsets provide a broad range of Internet services, e.g. an organiser, online-banking, e-mail, entertainment and infotainment services or games.

The Portal and Content

In launching iMode, NTT DoCoMo has established a portal site and lined up various content providers accessible for users directly from iMode's menu bar. NTT developed a flexible billing method to charge a commission for the services rendered by its first tier. Subscribers can browse non-official websites as well, but only official content providers can charge a monthly fee for a service via NTT DoCoMo. iMode's critical success factors are low costs due to the charging of data volume and not time, rapid continuous access to the Internet and an open platform strategy ensuring that consumers' preferred services and most popular sites get primary placement. The three most used services of iMode are: sending and receiving e-mail, surfing on the Internet and downloading data.

Users pay a fixed charge of $3 per month and about one cent per data packet (128 bytes). NTT DoCoMo generates cash in two ways: they charge users for the volume of data transmitted over their packet data network and they garner nine per cent of the revenue from services using their billing system.

The Technological Basis

Technically, iMode is an overlay over NTT-DoCoMo's ordinary mobile voice system with the only difference that iMode is principally "always on", provided that you are located in an area reached by the iMode signal.

In January 2001, the Dutch KPN Mobile N.V., TIM (Telecom Italia Mobile) and NTT DoCoMo entered into a three-party Memorandum of Understanding to establish a leading pan-European mobile multimedia platform by introducing iMode to the European market. As opposed to the situation in Japan, iMode in Europe will be based on GPRS (General Packet Radio Service) technology. A Joint Venture of KPN and DoCoMo with TIM as its partner, expects to launch its C-HTML-based Internet offerings in Europe by the end of 2001, initially targeting over 30 million mobile subscribers of KPN Mobile and TIM in Belgium, Germany, Italy and the Netherlands. The overall potential of iMode subscribers is about 160 million within the European market.

Main Content Aggregation Revenue Streams
Subscription fees for using the aggregator's portal (e.g. public TV and Premiere World).
Revenues for VoD/PPV services in order to pay for provided content. Depending on the role of content producers and distributors, these revenues can be paid directly to the aggregator or via a revenue sharing model.
Sales of content to other portals.
Revenues by transactions as present in all markets, explained in detail in chapter 5.1.3.

> Broadband access provision will become a commodity - in the near future bandwidth will cease to be a limiting factor. Without their own aggregator or content production, distributors will compete on price or be located in niches.

3.3   Content Distribution

Now, with content distribution, analyses focus on the third market of the Get Audience area. Based on the idea that multi-channel access portals offer wired and wireless access, this chapter deals with the diverse access technologies. It concentrates on their future perspectives, roles and different possibilities to access broadband content, describing players & competition, infrastructure, demand, advantages/disadvantages and the future scenario focussing on Austria, Switzerland and Germany. Other countries in Europe, even neighbouring ones, might show a different picture of the infrastructure situation.

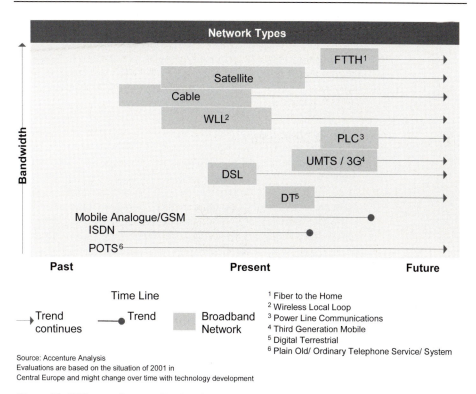

Network Types

- FTTH[1]
- Satellite
- Cable
- WLL[2]
- PLC[3]
- UMTS / 3G[4]
- DSL
- DT[5]
- Mobile Analogue/GSM
- ISDN
- POTS[6]

Bandwidth

Past   Present   Future

Time Line

→ Trend continues   ●— Trend   � Broadband Network

[1] Fiber to the Home
[2] Wireless Local Loop
[3] Power Line Communications
[4] Third Generation Mobile
[5] Digital Terrestrial
[6] Plain Old/ Ordinary Telephone Service/ System

Source: Accenture Analysis
Evaluations are based on the situation of 2001 in
Central Europe and might change over time with technology development

Chart 19: Different Access Technologies Will Penetrate the Market

The market for access technology is undergoing a period of change. In the Internet business, access via telephone line and modem has become a commodity and many people are now upgrading to broadband access. Furthermore, many of the newly emerging content types (as discussed in the previous chapters) will need broadband. This presents two obvious challenges for those intending to offer the whole service/content spectrum: bandwidth and a return channel. In this context, different access technologies are competing in terms of time to market and technological maturity (Chart 19, Chart 20). DSL and cable, in particular, are the recent stars and satellite will follow soon. The mobile world will see new standards like UMTS in the years to come. And within a few years, well-known terrestrial television will give way to interactive TV. Therefore, the competition will increase and other upcoming and slower technologies, e.g. Powerline Communication (PLC) or Wireless Local Loop (WLL), will have a hard time gaining market share in the mass market. They are seen as niche players in a special segment (e.g. home networks, SME).

As broadband capacity increases, players in the field of broadband distribution will not be able to rely only on revenues and huge profits from broadband access provisioning. As in the Internet access game, broadband access will become a commodity and margins will decline because of a high increase in capacity. The audience will be able to choose from a variety of access technologies, right at their personal discretion with regard to convenience and situational reasons. And, the fact has to be considered that customers, once they have chosen an access provider, have proved to be reluctant to change their provider (as the liberalisation of the telephone and energy market has shown). Hence, in all areas of the content distribution game, time to market is the key success factor - customer relationship is the next step in the effort to retain more customers.

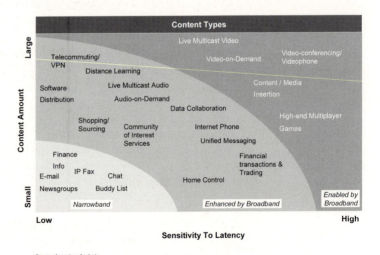

Source: Accenture Analysis

Chart 20: Service Types and their Requirements

3.3.1   Coax Cable

Using the TV cable infrastructure can offer fast access to the broadband world including high-speed Internet access and voice-over-IP-enabled telephony. In order to use cable for interactive and digital content and services, the cable network is currently being enhanced in terms of digitalisation and back-channel capability.

Players & competition: In addition to regional CATV operators of different sizes, UPC, Callahan, NTL and Klesch are the major European CATV operators, doing business in more than two European countries. In Germany - the largest European cable market - Callahan recently acquired 55% of the cable network of North-Rhine-Westphalia and has signed a letter of intent for a 55% stake in that of Baden-Württemberg with Deutsche Telekom. The other large player in Germany, the Klesch/Liberty consortium, has just signed a letter of intent to acquire a 55% stake in the cable network in the six regions of Bavaria, Berlin/Brandenburg, Hamburg/Schleswig-Holstein/Mecklenburg/ Vorpommern, Lower Saxony/Bremen, Rhineland-Palatinate/ Saarland and Saxony/Saxony-Anhalt/Thuringia with a 10-year option to acquire another 20% for the same amount. Klesch already holds a majority of the Hessen cable network.

In Switzerland, NTL bought the largest CATV operator Cablecom. About 300 CATV companies operate approx. 1,300 networks in Switzerland. Around 71% of them are private firms, 21% public authorities and 8% hybrid organisations. The market structure has become highly concentrated with Cablecom (now NTL) connecting 1.7 million households and thus controlling more than half of the Swiss CATV market.

In Austria, about a million homes are currently connected to broadband cable with UPC-owned Telekabel Wien being the market leader. UPC-owned Chello Broadband offers Internet access via cable using regional cable networks (e.g. Telekabel Wien or Telesystems Tirol).

Infrastructure: Digital bi-directional cable platforms are currently being deployed. Investments to upgrade the German network to bi-directionality are estimated to be around €3.5 billion for "Netzebene 4" and €7.5 billion for "Netzebene 2 and 3" (see Accenture study "Reinventing Cable TV Business", August 2000).

Demand: For cable services, demand will significantly rise over the next few years. In 2001 the revenue (for German cable operators in TV & residential telecommunications only) will exceed $2 billion. The fulfilment is highly dependent on the timely network upgrade and a common standard for termination devices (e.g. set-top boxes or cable modems).

Advantages:

- Last-mile technology
- High bandwidth
- High transmission quality
- Residential and business customers
- Larger customer base than satellite
- Key players gain economies of scale by implementing European rollout

Disadvantages:

- Bandwidth-sharing between users
- Time-to-market disadvantage compared to DSL
- Fragmented market structure leads to problems with fast service rollout
- Complex market structure due to separation in Germany between network levels 3 and 4 (Netzebene 3 and 4).
- Cable modem required (no standard modem)
- No single standard for set-top boxes
- High up-front investments

Future scenario: The EU and German governments are not happy with the solution as it is now, having Deutsche Telekom removed as a monopoly but Klesch/Liberty and Callahan as a duopoly and Deutsche Telekom with a 45% minority stake.

3.3.2   Satellite

The current generation of satellite systems was primarily designed for broadcast and offers digital one-way high-bandwidth transmission. Using different technical options as a narrowband return path, two-way communication has been established as a bridge to the real two-way broadband communication. The new generation of satellite systems will offer two-way, asymmetric broadband/multicast capabilities. However, real two-way symmetric broadband/point-to-point communication capabilities are planned for 2003/5 at the earliest.

Players & competition: Currently, the market is still dominated by broadcast systems. The majority of satellite homes use their set-top box mainly for free-to-air reception. Operating twelve satellites, SES Astra offers the leading direct-to-the-home satellite system in Europe. The pay-TV market is comparably small, with Premiere World (pay-TV channel) dominating the German satellite pay-TV market.

There are companies offering Internet access via satellite, too. In Germany, for example, Strato AG offers connections from 1.6 to 4 Mbps downstream via satellite using the telephone line for upstream connections. At this stage, however, it is not possible to use the same satellite connection for TV and Internet purposes. This planned feature offered at an acceptable price range will represent a big additional benefit for the customer. Some services such as videoconferencing are already available via satellite for businesses, but too expensive for private users.

Infrastructure: In the past, satellites have traditionally been used for TV broadcasting and as telephony backbone. More and more innovative services are now emerging in the narrowband (e.g. data services such as e-mail or paging, location-based services, telematic) and broadband (e.g. Internet access, e-commerce, tele-health etc.) areas. Out of these different groups of satellite services, multimedia satellite services allow satellite operators to enter the broadband market. For example, SES Astra is currently in the process of launching its broadband interactive system allowing data rates of up to 2 Mbits/s upstream in the KA-band and up to 38 Mbits/s downstream in the KU-band.

Broadcast services:

- Fixed satellite services are the traditional satellite services for broadcast providers. The main operators include EutelSat, SES Astra, PanAmSat, Loral Skynet, IntelSat/ComSat, GE Americom, AsiaSat, NSAB and StarOne.

- Direct-to-the-home systems that provide homes with hundreds of satellite channels for digital TV and audio entertainment supporting the "pay per view" logic. The main operators include Astra, DirectTV, PrimeStar, Echostar, USSB, Pegasus, BskyB, Canal+, Sky LA, Indovision and Sky Perfect TV.

- Digital audio radio services use satellite communications in order to provide advanced digital audio radio services, offering more than 50 channels. The main companies are CD Radio, American Mobile Radio Corporation and WorldSpace.

Telecom services:

- LEOs (low earth orbits) offer a large set of low-cost data and message transmission services. The main providers are ORBCOMM, Leo One USA, Final Analysis and ESAT.

- Satellite Imaging offers specific services for topographic imaging. The main operators are SpaceImaging and OrbImage.

- Multimedia Satellite Services offer bandwidth on demand and are designed to support interactive broadband services e.g. Internet access, videoconferencing and intranets. They mainly operate in LEO/MEO (low/medium earth orbits).

Future operators are Teledesic or SkyBridge. However, GEO systems are also staking their claim, e.g. Astrolink, Spaceway, EuroSkyWay and West.

- Mobile satellite services are mobile telephony services that cover regional areas or the world as a whole. The worldwide Iridium system was the only attempted operating system to date. Future regional systems will include Asian Cellular Satellite (AceS), Thuraya and Alcatel. Future global systems will include Globstar, ICO, Ellipso and Constellation.

Demand: Two-way broadband communication is not yet offered in an acceptable price range. Demand comes mainly from broadcast applications (satellite TV). Beside pure TV broadcasting there will be high demand for digital broadcast applications (transmitting digital files to several locations at a cost advantage compared to physical transport of CD-ROMs, tapes, etc.), satellite systems used as backbone backup solutions and combinations of TV-broadcasting and Internet activities where the user can shop or access information about TV movies, etc.

Advantages:

- Last-mile technology
- Perfect coverage (nearly 100% coverage of the earth even in exotic areas and maritime enterprises, e.g. Indian software production depends on satellite infrastructure for data exchange)
- Backbone technology (due to high transmission rates)
- Innovative services (e.g. traffic telematics, object security, ecology monitoring)

Disadvantages:

- Bandwidth-sharing between users
- Long time-to-market for interactive services
- Long transmission delay (GEOs, geo-synchronous earth orbits)
- Transmission quality affected by bad weather

- Limited downstream bandwidth for current satellite system generation (due to limited transponder capacity)
- High up-front investments
- Expensive satellite terminals

Future scenario: The new satellite generation feasible for interactive services is expected to be launched soon. Gilat Satellite Networks recently announced an initiative aiming to install 100,000 very small aperture terminals (VSAT). The planned launch within the third quarter of 2001 would make it the first two-way broadband satellite service available in Europe. Astra is also planning to launch its broadband interactive system in the third quarter of 2001. Teledesic is investing several billion dollars to offer broadband via its planned 288 satellites ("Internet in the Sky") starting in the year 2003.

However, applications relying on high bandwidth face a problem if transmitted via satellite due to bandwidth-sharing and limited transponder capacities. Also, due to the late market entry as interactive broadband last-mile technology, we expect this technology to be too late for the residential broadband mass market or as an end-to-end transmission technology. Nevertheless, satellite services will show an impressive market growth in the area of backbone/backup systems and digital broadcast applications. Bundles of satellite services with other access technologies (hybrid systems) will continue to be another niche market. Additionally, in terms of interactive broadband services, only some specific and appropriate (economical, technical) services are expected to be offered via satellite systems (e. g. B2B services, security- or ecology-monitoring and internet access).

3.3.3   **Digital Terrestrial**

Using digital terrestrial technology - mainly known for multi-channel TV and radio, - idTV and interactive services (full Internet access) could be delivered to the home through an ordinary TV aerial (no dish or cable) and a back channel. This requires a decoder (set-top box) or a digital TV that has the technology already built in. Digital Terrestrial in Germany, Austria and Switzerland is currently in the planning phase.

Players & competition: Trials have already been launched in Germany in spring 1999. NDR (Norddeutscher Rundfunk) began trials in Lower Saxony, Brunswick, Bremen, Hamburg and Hanover, carrying channels from the public broadcasters ARD and ZDF, whereas Deutsche Telekom launched trials in Berlin-Brandenburg and Hanover with the programming of commercial broadcasters. Trials will be expanded to other regions in the near future in order to prepare the commercial rollout planned for the beginning of 2002. Deutsche Telekom will operate the distribution infrastructure. Due to market entry barriers, no new market players are expected. The established analogue terrestrial companies (e.g. ARD and ZDF) are expected to provide this kind of service.

In the UK, ONdigital (a joint venture between Carlton Communications and Granada, two of the existing independent interactive television operators on analogue terrestrial television) launched a commercial digital product as early as four years ago. The main competitor for Ondigital is satellite-based BskyB.

In contrast to countries like the UK, Spain, Japan and Australia, where digital terrestrial will play a major role as a local-loop technology, for Germany only a significant role in the broadcast area is expected.

Infrastructure: Digital terrestrial TV will raise the number of regional and special-interest programs broadcast. The pictures have very high quality, the sound is crystal clear and it is possible to watch in cinema-style wide screen format (using the latest television sets).

Demand: Forecasts predict the Digital Terrestrial penetration in Germany, Austria and Switzerland with significant growth starting in 2002 although the total number of digital terrestrial homes tends to stay small in relation to other broadband access technologies.

Advantages:

- Only digital set-top boxes or TVs with built-in functions are needed

Disadvantages:

- New quality TV needed
- Investment in new transmission and compression technology
- No return path directly provided, no direct interactivity
- Long time-to-market

Future scenario: Due to the late rollout of digital terrestrial TV, this technology will not play a significant role as a digital TV platform in Germany. Traditional terrestrial distribution will develop into a niche market in Germany in the near future, since the majority of viewers will have already migrated to DSL, cable and satellite services.

3.3.4   xDSL

Digital subscriber line ("x" represents different technology options) is a technology for transporting data via a twisted-pair copper wire from the local exchange to business premises or homes. Two-way communication can be established using the telephone line. At the current stage of development, DSL offers a bandwidth of up to 2 Mbps in synchronous mode (SDSL) and 8 Mbps in asynchronous mode (ADSL).

Players & competition: xDSL is offered by nearly every telco. There are at least 40 vendors for DSL technology worldwide. In Germany, Deutsche Telekom launched its business service in April 1999 and its residential service in July 1999. Due to regulations, market entry barriers for service providers in

Germany are low. More than 40 DSL service providers (e.g. Deutsche Telekom, Arcor, Colt, QS Communications, KPN Quest and MobilCom) already compete in Germany.

Today the market is dominated by the "T-DSL" product of Deutsche Telekom, although it offers only 768 Kbps downstream and 128 Kbps upstream. Competition has not really taken hold and is currently being discussed by the regulation authorities. Other carriers, for example QSC, offer symmetrical DSL and are ready to gain market share. The market is extremely price competitive and a shakeout is expected very soon.

In Switzerland, Swisscom only offers wholesale packages (delivering 256 Kbps downstream and 64 Kbps upstream or 512/128 Kbps) for other service providers (currently Easynet, Econophone VTX and Swisscom subsidiary Bluewin).

In Austria, Telekom-Austria's subsidiary A-Online was the first in late 1999 to offer ADSL delivering 256 Kbps (upstream). By September 2000 they had a theoretical range of 75% of Austria and 28,000 ADSL customers. Around 20 providers now offer ADSL services.

Infrastructure: Running on the telephone cable, the present network is already installed but data transportation in the backbone has to be enhanced. Depending on the location, equipment for DSL access has to be installed. There are many variations of xDSL (the most important are HDSL, SDSL, UDSL, ADSL, VDSL, of which some are still in trial). The main differences are the proportion between fibre and copper, the transmission symmetry (symmetric vs. asymmetric) and the transmission speed. "Fast Internet Access" (2 Mbps) is currently offered, but in the near future - using VDSL as a "download pipe"- up to 26 Mbps can be offered depending on the quality of the existing copper and the distance between the main distribution frame and the customer premises. This requires an extension of the fibre network to the curb.

Demand: Especially for residential customers, SoHos (small offices, home offices) and SMEs, xDSL offers an interesting broadband access technology. In terms of revenues, residential customers will represent the most interesting customer segment. Due to the high number of PC-based Internet users and strong promotion of the DSL product, there is a high demand for xDSL, resulting in a delay for installation (in the case of T-DSL, this can be up to 7 months).

Advantages:

- Time-to-market
- No bandwidth-sharing (guaranteed bandwidth)
- Theoretical access to all 37 million households in Germany; high coverage
- No need for telco investment in copper infrastructure
- Lower costs for upgrading networks ($300-400 per subscriber in comparison to $600-750 per subscriber for cable)
- Parallel transport of voice and data along one line (voice: analogue or ISDN – data: ATM, Frame Relay, Ethernet or IP)
- Secure point-to-point connection in comparison to the tree-structure of cable networks requiring encryption

Disadvantages:

- Market in Germany dominated by Deutsche Telekom, rest of market fragmented
- DSL comprises many variations of different technologies and standards
- Backbone network to main distribution switches needs to be upgraded. At present only a portion of potential customers are able to switch to xDSL because of backhaul bottleneck
- Performance is distance-sensitive (maximal transmission speed up to a distance of 3km)

Future scenario: It is expected that DSL will develop faster than other broadband technologies in the next few years due to coverage and ease of installation. DSL is the main competitor for cable in the future and demands only limited investment in the infrastructure. Break-even should be reached earlier. Currently, xDSL is very attractive in Germany and will start to penetrate the market in the urban areas (due to range restrictions).

3.3.5   UMTS / 3G Mobile

The old digital mobile platform GSM (2nd generation) is currently being upgraded to GPRS, EDGE or HSCSD (all $2^{1/2}$ generation). The next quantum leap will be the rollout of UMTS (3rd generation), capable of delivering broadband services via wireless terminals.

Players & competition: In Germany, GPRS is already offered by the existing GSM players. Following the sale of licenses for nearly €50 billion, there are six UMTS players on the market. In addition to the GSM network players, T-Mobil, D2Vodafone, E-Plus and Viag Interkom, Mobilcom and Group 3G also compete on the new UMTS market.

In Switzerland, there are three GSM players - Swisscom, Diax and Orange - whereas UMTS licenses have been bought by Swisscom, dSpeed, Orange and Team 3G for a total of 205 million Swiss francs.

In Austria, the six UMTS license winners are Connect, Hutchison 3G, Mannesmann 3G, max.mobil, Mobilkom and 3G Mobile.

Infrastructure: Today, GSM networks are ideal for transmitting digital voice, whereas the next generation of networks will optimise data communication. Therefore, wireless networks are on the move to packet-based transmission in order to use the increasing data speed and capacity more efficiently. For UMTS, an entire new network infrastructure has to be built up, whereas the new $2^{1/2}$G technology will use the existing GSM network. GPRS (General Packet Switched Radio Service) is a packet-switched system using GSM and/or TDMA circuits but shares capacity dynamically between users to maximise channel utilisation efficiency.

HSCSD offers up to 56 Kbps (the German mobile provider E-Plus currently offers 43 Kbps downstream, 14 Kbps upstream or 28 Kbps both ways), while GPRS offers up to 170 Kbps (first tests show data rates below HSCSD) and EDGE up to 384 Kbps. The UMTS transmission speed will be between 512 Kbps (mobile) and 2 Mbps (stationary). Initial trials show that the UMTS-technology will probably not achieve the maximum bandwidth in the short run.

Demand: The demand for mobile broadband services will focus on m-commerce (e.g. personal finance, advertising, shopping, online gaming and mobile payment), video or mp3 download and location-based services (i.e. finding, tracking, etc.). The Japanese carrier NTT DoCoMo will offer video transfer using the new mp4 protocol. This alliance of NTT DoCoMo, KPN and ePlus plans to introduce "iMode", a portable Internet solution, on the European market.

We expect a more transaction-oriented personal behaviour of the users in contrast to the browsing/surfing behaviour we find in the wireline area. In the B2B arena, extranet access or ASP (Application Service Providing) will dominate the market place. UMTS will allow M2M communications with high bandwidth.

Advantages:

- High transmission rate
- Mobility as a differentiator from other access technologies
- Location information can be accessed to offer location-based services
- Trendy product and status symbol
- "Always on" mode with UMTS and GPRS

Disadvantages:

- High UMTS license costs lead to high market risk for carriers
- High investment costs for build-up of network (UMTS) or upgrade (EDGE, GPRS)
- Transmission speed not comparable to fixed broadband services

Future scenario: HSCSD, GPRS and EDGE will be available in 2001 offering interactive data services and Internet access to users, but no broadband yet. Some services are already available, but carriers hesitate with marketing due to unsure network stability. UMTS is expected to be launched in 2003/2004. There will be many more broadband-intensive and interactive services running on UMTS than on $2^{1/2}$G networks. Cannibalisation effects between UMTS, 2G and $2^{1/2}$G networks will arise, resulting in a mix between UMTS in the urban areas and $2^{1/2}$G services in the rural areas. Although the request for applications in the data communication arena will increase over the next few years, voice will still be the most important application.

The extremely high upfront investments are resulting in several issues. On the one hand, companies are looking for strong partnerships to gain a competitive advantage, as can be seen by Deutsche Telekom and British Telecom working together on building 3G networks in Britain and Germany. This co-operation helps to improve the business case by saving costs (expected to amount to billions of euros), because networks can be built more quickly and it will be possible to reduce the number of base stations and masts needed for transmission. On the other hand, there will be strong pressure for the network players to migrate to content and additional value-added services or to enter the m-commerce arena, where they find higher margins. Over the next few years, the major task for the operators will be to acquire other companies or to form partnerships with companies offering attractive content and services, as well as to build up a wireless portal to dominate end-user access.

Due to the fact that there are no real alternatives (mobile satellite services still have a very low penetration), UMTS has a clear competitive advantage when it comes to mobile data access and will dominate this market. UMTS could offer a wireless local loop for residential customers and SoHos with modest bandwidth requirements. Due to relatively low bandwidth, UMTS will not dominate the wireless local loop market for SMEs, but can be seen as a complementary technology option.

3.3.6	**Wireless Local Loop (WLL)**

Wireless local loop services have already been launched. This technology offers wireless access to broadband bandwidth without depending on cable or fibre. The point-to-multipoint service runs in two different frequency bands: WLL on the 3.5 GHz band has a large coverage (up to 10km) with a slightly smaller amount of bandwidth available per customer. The service on the 26 GHz band can cover a radius of up to 3.5km with a bandwidth per customer of up to 12 Mbps. In addition, point-to-point radio links are used to connect two sites with a transmission speed of up to 155 Mbps or as a connection to the backhaul.

Players & competition: The German market was the first broadband WLL allocation in Europe. 12 companies were awarded licenses that were allocated by means of a "beauty contest" in August 1999. We can identify the pure-play WLL companies like FirstMark, Star 21 Networks or Broadnet. On the other hand there are multi-carriers like VIAG Interkom, UPC, Tele2 or Arcor. Altogether there are around 20 players in Germany. Licenses are regional and not nation-wide, which leads to a fragmentised geographic coverage by every licensee. Many licensing areas and several licensees in one area lead to a more complex go-to-market strategy. The establishment of European growth strategy is difficult because of different licensing procedures in the EU. However, European coverage is especially targeted by FirstMark and BroadNet.

In Switzerland, WLL licenses have been sold for 582 million Swiss francs (of which FirstMark and UPC alone paid around 59%). The nine winners are FirstMark, UPC, Europe i Switzerland, Callino, BroadNet, Sunrise, Star 21 Networks, Telecom Ventures of Switzerland and Commcare. Beside the WLL niche players, only Sunrise and Commcare can be considered to be established Swiss telecommunications companies. Initial WLL services are expected to be offered in 2001.

In Austria, the applicants for a license are BroadNet, Callino, European Telecom International, max.mobil Telekommunikation Service, Mobilkom Austria, Star 21 Networks and Tele2 Telecommunication Services.

Infrastructure: The WLL infrastructure consists of a base station (antennas on a high building and a small router) and a smaller antenna with a router interface at the customer site. Per base station, a total capacity of up to 144 Mbps is feasible. This is distributed to the customer premise equipment, which can handle up to 12 Mbps. Similar to a main distribution frame of the wire-line technology, the base station needs to be connected to a backbone. The WLL infrastructure is currently under construction and can at this stage only be used in a few areas. Lucent, Siemens, Ericsson, Ascom and Nortel are some of the companies offering the infrastructure for WLL.

Demand: At the moment, most providers only have pilot customers. Due to the lack of customer base, the rollout is very slow and it will take longer to realise the business case. The early advantage of quick installation time is under discussion, as securing roof rights for the base stations and building these with scarce resources is taking longer than expected. Customers do not have any experience with this technology and have a negative awareness about the quality. Nevertheless, the time-to-market is still attractive and customer segments such as small and medium enterprises are an attractive target.

Advantages:

- Offering local loop without significant construction activities
- Broadband connection with high quality within 3.5km range
- Flexible bandwidth, bandwidth on demand

- Fast time-to-market and short time for installation
- Price is more attractive than fibre

Disadvantages:

- Currently high infrastructure investment up-front ($100,000-200,000 per base station and $2,500 for the customer terminal and costly site acquisition), but costs are expected to decrease
- Low coverage (license areas, line-of-sight limitation)
- Influenced by weather
- Resistance against antennas (problem for site acquisition and base stations)
- Maintenance very difficult due to necessary site access (e.g. roof tops)
- No pan-European services possible
- No Germany-wide service offerings

Future scenario: Target customers are mostly SMEs that do not use a direct fibre access (FTTH) up to now because of cost considerations or due to their location. With respect to the target customers, data services (leased lines, Internet access, IP-VPN) and net sourcing services (hosting/co-location, application services), complemented by voice services (switchless voice, VoIP), are expected to become the primary offerings.

With prices comparable to city-carrier services, a strong focus will be on cities and urban areas and moving to more rural areas in the next 2-3 years.

Currently, most WLL providers encounter problems with their rollout and operations. Hence, the growth rate is not as fast as predicted. There is a danger that the WLL companies might become acquisition targets for full-service carriers or carriers with other access technology (e.g. fibre, xDSL) as well as new UMTS entrants who need a backbone and backhaul infrastructure for their mobile phone. Therefore, some WLL providers will complete their market offering by building up additional access technologies.

3.3.7   Power Line Communications (PLC)

Powerline is the most uncertain access technology. While companies like Siemens and Bewag dropped PLC, utilities like EnBW, RWE and MVV are preparing to launch such services. Although technology breakthrough has been predicted for years, many legal and technical issues are still under discussion. Nevertheless, the concept has a significant long-term potential as soon as these issues are solved.

Players & competition: PLC presents an attractive opportunity for utility companies to enter the telecommunications market. We expect PLC services to be offered by large utilities e.g. RWE, GEW Köln, EnBW and E.ON (through subsidiaries Oneline and AVACON), although the first hype of PLC has died down and several leading companies in the PLC field have withdrawn (e.g. Nor.web, a joint-venture between Nortel Networks and United Utilities). Several trials are currently running (e.g. EnBW or Alcatel together with MVV Mannheim).

Ascom has partnered with 14 companies and started eleven field trials, in Switzerland together with the telecommunications provider Diax. In Austria, power supplier ENV is running a pilot rollout with 20 customers for Internet access via power lines together with Ascom. Commercial rollout is planned by late 2001 with an initial capacity of 1.3 Mbps and up to 10 Mbps in the final stage.

Infrastructure: Symmetrical broadband telecommunications traffic is carried over existing power electric distribution infrastructure. However, PLC requires additional infrastructure at the electricity transformer in order to connect the power network with a data switch. Trials for PLC currently offer bandwidth up to 3 Mbps that is shared by households connected to electricity transformers. For in-house networks, around 10 Mbps are targeted. Munich-based Polytrax Information AG and Hitachi have introduced a modem with a maximum speed of 2.4 Mbit/s, able to provide 1.5Mbit/s in networks with heavy load.

Demand: At the moment, the most attractive market for PLC seems to be the home networking area. Home networking means the connection of devices within the home for the purpose of transporting and sharing information. This can include information-sharing between PCs, audio systems, video systems etc. In addition, new value-added services like monitoring, security services and direct reordering from machines will come to the attention of customers. Any home appliance (e.g. monitoring, controlling and security) that has the right chipset can be connected to the network, giving seemingly extensive capability to home networking. Internet access and telephony (e.g. baby phones are early products that have already been using similar technology for a long time) can be shared among several devices within the home.

Advantages:

- Potential of 35 million utility customers constitutes a very attractive customer base
- Technology for the last mile
- Network enhancement without major investments for utilities
- Low installation costs and fast installation time as soon as the technology is stable
- Supporting innovative services in house networks
- Always-on mode

Disadvantages

- Bandwidth-sharing between households
- Timing of rollout delayed
- Susceptible to interference (technology still being tested)
- Electromagnetic tolerance / sociability
- Maximum transmission distance of 350m before a repeater is needed
- Lack of global standards
- Telecommunications regulations are unclear regarding power-line services all over Europe (e.g. case-by-case regional permissions for trials)

Future scenario: Some utilities (e.g. RWE and EnBW) have announced the earliest possible rollout date to be late 2001. PLC is mainly attractive for residential customers, although it provides a backup opportunity for SoHos and business customers. Right now, Power Plus Communications (PPC), a subsidiary of MVV Energie AG, is piloting Powerline with 200 households in Mannheim. The service is priced at 20% below Telekom ISDN at a speed of 30 kbit/s. PPC has already marketed PLC to five utilities in Germany and Austria, and is projecting 500,000 household customers by 2004.

In addition to offering Internet access, telephony and broadband transport, PLC will be established in the areas of narrowband remote maintenance and home networking. In this market niche, PLC will be bundled together with other complementary access technologies. PLC will trigger the demand for home networking (and facility management) and could be the surprising "joker" in the broadband arena.

3.3.8

Fibre To The Home (FTTH)

Fibre to the home offers potentially very high bandwidth and, once in the ground, fibre access mechanisms can be upgraded by adding new equipment lighting the fibre. The novelty about FTTH is that this access technology is offered not only for larger business customers but also for residential customers, SoHos and SMEs.

Players & competition: In Europe, FTTH is just at the beginning of its life cycle. In Germany, in particular, the Berlin-located carrier BerliKomm is just preparing to offer FTTH, Bredbandsbolaget is a leading FTTH player in Sweden, and NTL has bought a 25% stake to advance its own international broadband strategy.

Most of the world's largest telcos are limiting their actual FTTH deployment to small-scale trials. In the US, many smaller carriers and newcomers are eagerly pushing ahead with actual product offerings and are creating a significant base of fibre users. Especially US BellSouth is playing an active role. Other players in the US are Integrated Broadband Networks (IBN), CoreComm, ClearWorks and Rye Telephone.

Infrastructure: Fibre to the home (FTTH) is a network where an optical fibre runs from the telephone switch to the subscriber's premises. Metropolitan area networks (MANs) are at the heart of fibre deployment, with the installation of fibre rings from which fibre may be dropped to customer sites. Until 1988, a single fibre was needed to connect to each home separately. Recent advances permit N-way distribution to many homes through resource-sharing circuits.

The progress towards fibre to the home (FTTH) can be achieved via a series of intermediate steps using a combination of fibre and DSL technology. There is a progression from fibre to the remote (FFTR) through fibre to the cabinet (FTTCab) and fibre to the curb (FTTC). Fibre to the home (FTTH) describes the access architecture in which no copper is used at all in the operator's network. Bandwidth with fibre can be up to 652 Mbps. Together with satellite, fibre can offer the highest bandwidth for broadband access.

Demand: Target customers are residential homes with an affinity for technology in the first run as well as SoHos and small enterprises with a focus in the multimedia area.

Advantages:

- High transmission speed
- No upgrades for old networks needed
- Low operational costs (maintenance, provisioning, facility planning)

Disadvantages:

- High cost and time for deployment (current new techniques for building the network reduce the infrastructure costs significantly)
- Not ready for commercial launch yet (time-to-market)
- In the process of standardisation

Future scenario: A broad deployment of FTTH in Europe or Germany is not expected in the near future. Competition through existing infrastructure (cable, copper) will be too tough to allow fast-track penetration of the residential market. FTTH offers an unmatched broadband access capacity, but due to high installation cost it will be a niche product in the next few years, offered to high-load multimedia business users.

3.3.9 Access Technology Overview and Trends

To bring all access technologies in one picture a summary of the different criteria to compare the access technologies is shown in Chart 21. The criteria and the evaluation are based on the current situation and subject to change as the technology develops.

	Cable	Satellite	Terrestrial	xDSL	UMTS	WLL	PLC	FTTH
Effective bandwidth capacity*	very high	very high	low	medium	low	high	low	very high
Mobility/Portability	very low	medium	medium	very low	very high	medium	low	very low
Ease of installation (end user)	low	low	high	high	very high	low	medium	low
Return channel capacity	high	low	very low	medium	low	medium	low	very high
Time to market (Availability)	medium	low	medium	high	very low	medium	low	low
Coverage	high	very high	high	very high	medium	low	very high	medium
Number of potential services**	very high	high	low	medium	medium	medium	medium	very high
Technology maturity	high	low	medium	high	low	medium	low	medium
Transmission quality	very high	high	low	medium	medium	low	low	very high
Network investment needed	high	very high	medium	low	very high	high	medium	high

*= Maximum transmission capacity taking into account limitations by sharing capacity between the number of users (Cable, satellite, UMTS, PLC, FTTH)

** = Possibility of offering the complete range of services (Telephone, TV, Internet,...)

○	◔	◑	◕	●
very low	low	medium	high	very high

Evaluations are based on the situation of 2001 in Central Europe and might change over time with technology development

Source: Accenture Analysis

Chart 21: Overview of Access Technologies

Basically, time-to-market turns out to be the critical success factor to reach the mass market. Combined with the inability of some access technologies to provide sufficient bandwidth (e.g. UMTS), to sustain the necessary transmission quality (e.g. WLL, PLC) or to offer a backward channel (terrestrial) in the near future, the overview shows which technology has the potential to become a preferred gateway (see Chart 21). Cable and xDSL are in a particularly good position to conquer the mass market in the short term.

Generally, competition will increase and bandwidth will not be a limiting factor in the future. Once the new market is saturated, not every technology will offer every service to every customer segment because of different success factors. The opportunity to offer bundled packages of idTV, Internet, voice and data will be another separating factor. Hence, different business strategies will appear due to different competitive advantages. Cable, terrestrial and DSL will mainly be seen in the private sector, while WLL and FTTH will probably concentrate on businesses. Satellite will be mainly used as a backbone technology and in hybrid systems. Also, it has the potential to offer specific services to the business and private sector. The future of PLC is still uncertain but its short-term potential should be in the home network area. UMTS will be a technology that, due to its mobility, will be adopted by businesses and households. Coverage could be a decisive factor when it comes to conquering niche markets.

With time-to-market being the main critical factor for success, broadband access via DSL and cable will dominate the mass markets. As demonstrated in Chart 22, forecasts show that DSL will have an advantage in Germany and Austria, while in Switzerland cable already has a dominant market position.

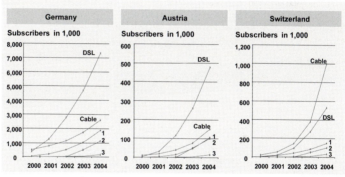

Source: IDC: DSL, cable and FWA; Ovum: 3rd generation network, Accenture Analysis
1 = WLL, 2 = UMTS, 3 = Terrestrial, Satellite, PLC and FTTH
Forecast only refers to interactive broadband subscribers

Chart 22: Market Forecast for Access Technologies in ASG

Main Content Distribution Revenue Streams
Subscription fees paid by the audience. In a commodity market, these subscription fees just for getting access to transmission (e.g. cable, DSL, UMTS) will decline over time.
Service fees from content and portal owners, likely based on a revenue sharing model, will reimburse the distributors for providing the required bandwidth.
Revenues by transactions as present in all markets, explained in detail in chapter 5.1.3.

> Synchronising the form of targeted merchandising and advertising with the offered content and the gathered specific customer interests will complement interruptive marketing.

4   Sell Audience

Huge investments to offer and deliver content on the one hand and the end users' limited willingness to pay for content on the other hand cause a shifting of revenues from the Get Audience to the Sell Audience. As mentioned earlier, this is common practise in the media business. So, Sell Audience opportunities in the interactive broadband media are to a certain extent necessary in order to refinance Get Audience investments. Hence, we now analyse the three main markets in which additional revenues can be generated: customer insight, advertising and merchandising. Many companies in the media business are following this approach to generate additional revenues out of the Sell Audience side. Yet beside being excellent managers of these markets, anyone interested in implementing the ideas described in the following chapters has to bear in mind that these Sell Audience markets will only pay off if the basis – which means the product, service or brand - on the Get Audience side is at least in a good shape. This is the prerequisite for any Sell Audience revenues.

In addition to these Sell Audience activities, interactive broadband services are able to source revenues in another profitable realm: transactions. These opportunities will be addressed in chapter 5.1.2.

Some capabilities and assets a company has built up can be used for several Sell Audience activities. As an example, the need for building up a strong brand arises especially in content aggregation. This brand is the prerequisite for successful advertising or merchandising. This leads to the idea of generating additional revenues with an intelligent combination of activities in order to allocate the company's resources as efficiently as possible. In this chapter, however, customer insight, advertising and merchandising are considered individually to pave the way for an analysis of possible combinations discussed in chapter 5.

Customer insight is based on the audience's personal and usage data. In interactive broadband services, the interactive character raises vast opportunities for collecting more and better customer data than in any other traditional media business. The gained superior customer insight serves many business needs and has tremendous impact on the two other Sell Audience markets presented here.

For advertising, the targeting capabilities gain in importance. The general trend realised in marketing clearly leads from interruptive/mass marketing to targeting narrow customer segments, allowing companies to address the needs and preferences of the customers more effectively. Permission advertising, a special form of one-to-one marketing, develops the idea of establishing more effective relationships with potential customers.

The third Sell Audience market, merchandising, is distinguished from advertising by the interactive broadband players leaving their neutral position: whereas in advertising, the message delivered to the customers does not require any involvement of the advertising platform (besides providing access to the customers), sides are taken in merchandising, and the company's own brand is used to support sales of merchandised products.

On the one hand, customer insight is the enabler of effective targeted advertising and merchandising as well as the customer-centric production of attractive content. On the other hand, it is a revenue source in itself.

4.1 Customer Insight

The permanent process of collecting all relevant information about the audience in general and the individual customers in a systematic way leads to superior knowledge about the customer, referred to as customer insight. Also, customer insight is a key success factor in the interactive broadband business, because it helps both to attract and hold broad audiences and to exploit almost all opportunities to sell this audience more efficiently. In order to obtain customer insight, companies need access to relevant customer data. All this is common understanding. However, interactive broadband offers opportunities to collect more and better data than any other media business. Hence, the following sections focus on this issue and the various business opportunities, grouped into internal and external usage of customer insight.

Before discussing the importance of customer insight for content producers, aggregators and distributors in greater detail, Chart 23 provides an overview of the main implications. Taking various kinds of customer data as a starting point, customer insight concentrates on a better understanding of customers by transforming raw data into valuable information. This information allows you to truly understand the customer's needs, to identify relevant customer segments according to all kinds of criteria as well as to define the total value of individual customers. The benefits are manifold: on the Get Audience markets, the well-known customer needs can be addressed more directly by offering tailored content and services. Marketing activities aiming to permanently attract broader audiences are enhanced as well. On the Sell Audience side, customer insight can be used to push advertising and merchandising revenues by enabling targeted marketing. In addition to these internal usages of customer insight, several means of external usage are possible within the given legal restrictions. Another important characteristic of customer insight indicated in Chart 23 is the permanent refining of the data basis, making use of every customer contact. As broadband media becomes interactive, these contacts can establish closer relationships with the customers, gaining deeper and deeper customer insight as explained in the following section.

Obviously, the idea of collecting, storing and processing customer data raises the question of the "transparent customer." Certainly, not all customers are willing to share their personal data. Some companies, e.g. American Express, see added value for their customers in offering superior data privacy standards. Although heavily using customer insight internally, they guarantee that no personal data is given away to third parties. Another approach is followed by the NAI, the network advertising initiative. This co-operative group of network advertisers from the Internet realm (including Doubleclick, Adforce, 24/7 Media and others) allows users to "opt-out" of targeted advertising delivered by the NAI ad networks. The initiative gives users the chance to choose for themselves and helps to build up trust. However, it is not yet clear what standards will be established in the different countries and to what extent difficulties will be overcome. Awareness of data privacy problems is an important first step towards developing a pragmatic solution.

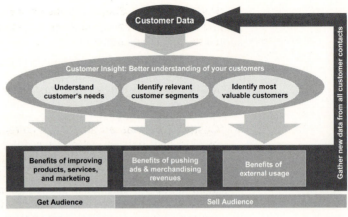

Source: Accenture Analysis

Chart 23: The Benefits of Customer Insight in the Interactive Broadband Business

How to Obtain Customer Insight

The ongoing process of customer insight can be divided into three separate stages: data collection and integration, data processing and generation of information with varied analytical methods, and finally the usage of this information for customer-insight-driven actions towards the customer. Chart 24 illustrates the three stages.

In the first stage, the collection and integration of data, access to customer data is obviously the starting point. In the interactive broadband media environment, there is a wide range of opportunities for gathering customer data:

- Registration: When registering to any broadband service, the user provides a basic set of demographic data, e.g. name, address and bank account.

- Usage tracking: Usage of content, communities, participation in permission marketing and other activities can be tracked in detail.

- Personalisation: Customers can personalise and configure their interfaces in the interactive broadband media, thereby giving away particularly reliable data on their preferences and interests.

- Partner data: Special transactions using encryption techniques, e.g. shopping transactions, can be integrated in co-operation with partners.

- Direct customer feedback: The ease of customer interaction allows you to ask the customer directly, using either simple questions like "did you like this film?" or more complex questionnaires and surveys.

- Customer data as an alternative currency: Customers can provide their data in order to pay for content or services, e.g. watching a video for free after having filled out a questionnaire. A striking example can be seen on today's Internet at bmw-films.com: Here, users have to provide their name, e-mail address and purchasing intentions in order to view short bmw movie commercials produced by famous movie directors.

- External data: Data collected elsewhere, e.g. market data, can be integrated.

Looking at these numerous data collection opportunities, two facts about the interactive broadband can lead to more and better data than is available elsewhere. Firstly, the intensity of usage makes a difference. Thinking about the extremely high penetration rate of analogue television of 98.6% and an average daily usage of 2-3 hours (example Germany) leaves no doubt about the great opportunities for collecting data when this medium becomes interactive. Combining Internet access with convenient always-on technology will add even more usage time. Secondly, the customer will become more involved than in today's media due to the interactive broadband's emotional character. Two implications of this emotional involvement can be distinguished. On the one hand, the customer will be more relaxed and willing to interact on the Internet than is the case today, where some kind of suspicion prevents many users from typing in relevant data. But, why not push some buttons on your remote control while relaxing watching a TV show? Or take part in a quiz when it's so easy? On the other hand, the data actually given by customers can be related to their emotional contexts, leading to new, more complex and valuable data.

Yet who in the interactive broadband business has access to the different data sources enumerated above? The content producers' data-gathering opportunities depend on the role of content aggregators and distributors. When a direct contact between the customer and producer is established, for example, from ordering special content directly, the producer can benefit from all of the above-mentioned ways of collecting customer data. If there is no such intensive communication between the producer and customer, e.g. when aggregators order content on a wholesale basis, or if content is produced for distributors offering content exclusively to their subscribers, the content producer's data access will be limited to direct customer feedback channels. On the other hand, the idea of data as an alternative currency could be extended: content aggregators and distributors for instance could pay for content with their customers' data. However, the media landscape suggests that every customer will utilise many more or less specialised content producers who can therefore only capture part of the full picture.

The content aggregator constitutes the customer interface with broad access to customer data and can use the full range of sources listed above. After having registered once, the customer can always be recognised via log-in. When proceeding to MCAPs, this procedure will even be independent of location and access channel - which is what makes MCAPs so valuable. With every customer using one content aggregator almost exclusively, the aggregators seem to be in a unique position to obtain customer data.

For the content distributors, it is easy to establish a stable client relationship, because every customer needs to subscribe in order to use the provided broadband access. The basic customer data is then already required for billing. Other more specific usage data is within reach, for example usage times or the amount of data flows. Depending on the technical details and the role of content aggregators and distributors in the established revenue model, it is even possible to monitor used content. If content or services are paid for on a revenue-sharing basis (as is the case for NTT DoCoMo's iMode), the distributors can easily record customer usage of their registered partners (e.g. using billing data), although the core business of content distributors (bandwidth for money) is not related to specific content or selection of specific content. A small problem might be that the customer will use more than one distributor to access interactive broadband services, thereby limiting the opportunities for distributors to gather data.

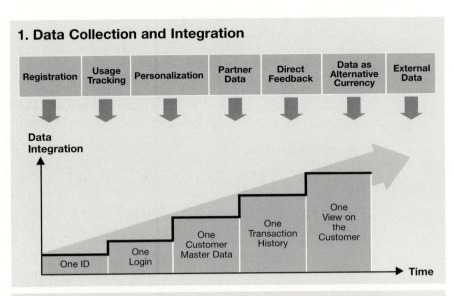

1. Data Collection and Integration

| Registration | Usage Tracking | Personalization | Partner Data | Direct Feedback | Data as Alternative Currency | External Data |

Data Integration

- One ID
- One Login
- One Customer Master Data
- One Transaction History
- One View on the Customer

Time

2. Data Processing and Information Generation/Editing

Data Mining, Data Analysis, Data Modelling

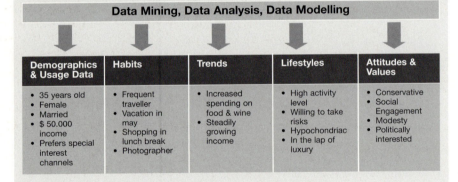

Demographics & Usage Data	Habits	Trends	Lifestyles	Attitudes & Values
• 35 years old • Female • Married • $ 50.000 income • Prefers special interest channels	• Frequent traveller • Vacation in may • Shopping in lunch break • Photographer	• Increased spending on food & wine • Steadily growing income	• High activity level • Willing to take risks • Hypochondriac • In the lap of luxury	• Conservative • Social Engagement • Modesty • Politically interested

3. Customer Insight-Driven Interactions

Targeted design and marketing of products and services
(product, place and time, promotion, and price)

Source: Accenture Analysis

Chart 24: The Three Stages of Customer Insight

Coming to the second step of the customer insight process, data processing and information generation, it is clear that the huge amount of data must be further analysed before it can become a basis for decision-making. Intelligent CRM systems and sophisticated data-mining tools are necessary (see excursion). The results of these efforts are very exhaustive profiles of the individual customers, as depicted in Chart 24. The many categories in which information can be organised include demographics, habits, trends, lifestyles and attitudes & values. The examples of information listed below the categories give an impression of the broad range of information available.

A company that already uses advanced types of categories to describe their customers is MTV. Positioned as a youth entertainment channel, they do not have much use for customer segmentation by age. Instead, MTV's approach is to define a number of different "mindsets", each of them describing a set of attitudes and lifestyle attributes. Every mindset is connected with specific preferences concerning music, clothes, sports and other topics.

The value of information generated by data processing differs significantly among the types of information. Obviously, the information about someone's specific habits and lifestyle, e.g. allowing you to conclude that this person is likely to book an adventure holiday within the next four weeks, is of much higher value than a person's age or sex. In general, the value of individual information corresponds to the difficulties of unveiling this information. Interactive broadband services are in a unique position to deliver the information that is most difficult to obtain, thereby emphasising the great potential of customer insight to add true value.

Moreover, interactive broadband services allow you to keep this valuable information up to date at all times. Compared to other available customer data, e.g. information sold by traditional listbrokers who tend to update their more detailed information with questionnaires only twice a year, the huge advantage of interactive broadband services is obvious. Realising the general trend in CRM towards using customer information almost instantly, this feature is of particular significance.

To sum it up, the second step in the customer insight process provides a complete set of information. It enables you to understand the customer's needs, to identify relevant customer segments and to define the customer's value.

The individual ability of content producers, aggregators and distributors to generate information is determined by their opportunities to collect data, already discussed in stage one. The third stage illustrated in Chart 24, customer-insight-driven interactions, will be addressed in the following two sections. However, the closed-loop effect of integrating new data gathered during these interactions should be mentioned here.

4.1.2   In-house Usage of Customer Insight

The three basic goals of customer insight are to understand the customer's needs, to identify relevant customer segments and to define the customer's value for a business, as illustrated in Chart 23. For interactive broadband services, all of these play a role when it comes to the benefits of using customer insight internally. Firstly, there are great opportunities on the Get Audience markets to improve products and services and to make marketing of these products and services more effective. Secondly, customer insight serves to increase advertising and merchandising revenues on the Sell Audience side.

Acquiring, Satisfying and Retaining Customers Efficiently

A true understanding of customer needs allows you to design tailor-made products and services. This capability is of key importance for ensuring satisfied and loyal customers as well as for winning new customers. Additionally, it increases retention and reduces churn. Costs can be saved significantly by streamlining operations to what really adds value to the customer. Besides, a thorough analysis of the customer's needs may help to elaborate additional services.

The ability to identify relevant customer segments is the enabler of targeted marketing. With detailed customer profiles and the right tools at hand, customers can easily be clustered according to all kinds of criteria. The different clusters can then be addressed individually following the idea of "delivering the right message to the right people". British Airways, for example, has analysed customer data to discover instances where an executive customer had flown one-way on British Airways but used another carrier on the return. It sent these valued customers a special mailing, headed "now we see you, now we don't," and offered a special incentive to use its services both ways. This approach can significantly save costs by minimising the advertising efforts wasted on customers who are simply not interested. Furthermore, the customer benefits from being exposed to less irrelevant and frustrating advertising (as will be discussed later in chapter 4.2).

Customer segmentation according to the individual customer's value is presented here as the third basic achievement of customer insight. This concept, often referred to as customer lifetime value, allows you to offer customers personalised interactive services as illustrated in Chart 25. This is of great strategic relevance when trying to ideally allocate the companies' resources. High-value customers are well worth a special service: an example is the "Vodafone D2 Platin Club", where mobile phone users generating especially high revenues can receive a second SIM card for free, buy a highly subsidised mobile phone every year, benefit from a better service and receive additional gifts, such free entrance to fair exhibitions, etc. These privileges are an effective way of keeping these high-value customers and reducing churn.

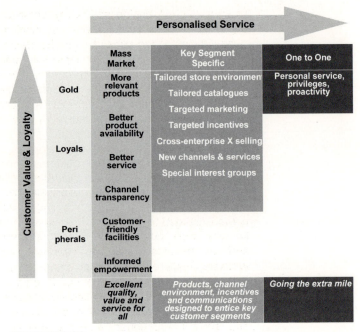

Source: Accenture Analysis

Chart 25: Customer Value and Personalised Service

Another major benefit of the in-house usage of customer insight is the improvement of the customer experience. Knowing the customers' needs, their specific value for the company and other customer characteristics allows you to better meet their expectations and raises great opportunities for up-selling and cross-selling additional interactive broadband services, for example in a call centre contact. Here, customers using only telephony and Internet access who have shown great interest in sports events may be willing to subscribe to a sports channel. The next time, they can receive extended personalised offerings, such as the full range of channels, thereby increasing sales and maximising every interaction's efficiency.

Increased Advertising & Merchandising Revenues

On the Sell Audience side, customer insight is the prerequisite for new revenues for interactive broadband services in addition to their original businesses. Only detailed knowledge about customers enables you to develop sophisticated, targeted forms of advertising and merchandising. The specific opportunities are discussed in greater detail in chapters 4.2 and 4.3. Here, we will only go into the main implications of customer insight for these markets. Thinking about the relationship between interactive broadband services and advertisers, customer insight will clearly strengthen the position of interactive broadband services.

Starting again with the aspect of understanding the customer's needs, it is interesting how advertisers can benefit from this knowledge. Clearly, an understanding of customer's needs is primarily related to advertisers themselves who are interested in knowing their own customers. When interactive broadband services are in a position to offer this understanding to advertisers, there are numerous opportunities for new services from which advertisers can benefit. One additional service is the prediction the of success of individual advertising campaigns. The interactivity of the broadband services allows you to track and record the response rates of earlier advertising efforts, for example by monitoring the use of "order now" buttons. In connection with the well-known customer needs, this forms an excellent basis for predicting the future behaviour of individual customers.

The ability to identify and address specific customer segments is, as mentioned above, important for every business. With respect to advertising and merchandising in interactive broadband services, several aspects must be highlighted. Firstly, it is necessary to be flexible in customer segmentation, because every advertiser targets different customer segments and prefers different forms of clustering. A company selling office supplies, for example, will be interested in the purchasing behaviour of a customer, while a recruiting company looking for the best talents would opt for psychographic segmentation instead. This flexibility implies high-performance CRM systems.

Secondly, the targeting capabilities can be turned into cash. For example, the CPM rates (cost per thousand impressions) of Internet banner advertising indicate that the chargeable prices depend heavily on targeting. On the Alta Vista home page, less than $15 is charged for a thousand impressions of a banner shown every time the site is run. When the same banner is positioned at a certain category, this CPM rate is $40, and at a sub-category even $60. Here, the targeting is realised via the accessed category, hopefully related to the customer's interest. The superior knowledge about the customer available for interactive broadband services allows much more specific and complex clustering. Moreover, Chart 26 depicts that advertisers are in fact willing to pay premium rates for these sophisticated clustering methods. The ability to address clusters grouped by psychographic characteristics like personality or lifestyle attributes ranks top, even higher than purchasing behaviour. These psychographic segmentation criteria are, like all others shown in the chart, within the scope of customer insight. Moreover, they are in fact almost exclusively available with the complex data analysis capabilities described in the previous section.

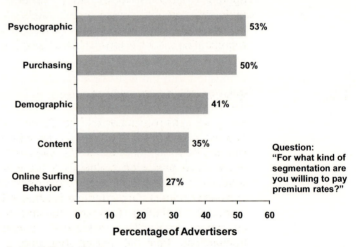

Source: Jupiter Executive Survey

Chart 26: Advertisers' Willingness to Pay Premium Rates for Clustered Target Groups

We shall conclude this section with some remarks on the individual ability of content producers, aggregators and distributors to use customer insight internally. Thinking about the improvements of products, services and marketing, most of the opportunities discussed hold true for all of these interactive broadband players. Being aware of customer needs is important for every company, no matter from what perspective the customers are addressed. Content producers can tailor the produced content, content aggregators can refine their way of selecting and presenting content and content distributors can think about the best tariffs, to name just a few examples. For content distributors, the opportunity to offer additional products and services will be especially important when their profits decline because their core service becomes a commodity. The existing customer relationship can lay the grounds for extended offerings, for example selling tickets. Pushing advertising and merchandising revenues is a hot topic, not only for content aggregators but also for content producers. They can use their understanding of customers' needs and segmentation capabilities to optimise their product placement activities representing a special form of advertising & merchandising.

4.1.3   External Usage of Customer Insight

Many successful companies like Beenz.com, E-centives, Inc. and Abacus Direct (acquired by Doubleclick in 1999) have realised that customer insight is a value in itself. Their business models are based on building up superior customer insight, which puts them in a position to offer several services to their business partners. One example is Abacus Direct, a US-based database marketing company running the world's largest customer transaction database, containing 3.5 billion transactions from 90 million households. Data has been acquired by bringing together transaction databases of more than 1,900 allied companies. Abacus Direct's main service is to provide mailing lists for their alliance members, making use of sophisticated data processing to compile these lists. First, the allied companies' proprietary customer database is analysed to extract the characteristics of the most valuable customers (data mining). Comparing these characteristics with the complete multi-merchant database allows you to evaluate the opportunities for selling products to individual customers and to extract the desired mailing list. One major catalogue retailer, for example, enjoyed an 80 per cent increase in average order size when mailing to a list filtered through Abacus Direct's database. The success of Abacus Direct is depicted in Chart 27. Another example, German Cocus AG, allows customers to provide their detailed profiles to be sold on a revenue-sharing basis. The customers get 40% of the sales revenues for their profiles. Although there are many more companies focusing on customer insight of which not all are as successful as the examples mentioned, there is evidence of the existence of a market for customer insight.

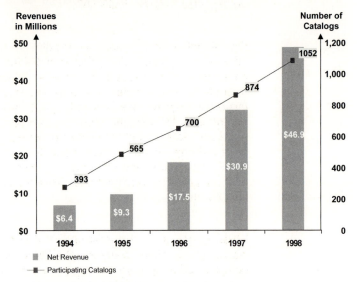

Source: Abacus Direct Financial Statements, Jupiter

Chart 27: The Success of Abacus Direct

Returning to interactive broadband media and their vast opportunities for building up customer insight, two major methods of external usage can be distinguished. On the one hand, there are opportunities for selling customer data directly. In Germany, where the legislation is particularly restrictive, selling of customer address lists accounts for DM 600 million in revenues each year. These addresses with additional information owned by specialised companies like Schober or AZ Bertelsmann, compared to the true customer insights discussed here, are quite primitive. Most addresses are enriched with only vague additional information, such as the customer's age approximated via the likelihood of their first name in the respective period. More detailed information is obtained through questionnaires, always facing the problem of not being up-to-date. The average prices paid for the right to use these addresses once vary from about DM 0.30 for simple information to up to DM 5 for more detailed lifestyle profiles. This broad range again shows the added value of high-quality data. When the detailed profiles that interactive broadband services can obtain enter the customer data market, the market volume can be expected to grow significantly.

On the other hand, customer insight can be shared among different companies. Co-operation between several partnering companies often focuses on strengthening customer loyalty through specific bonus programs. In addition to improved customer loyalty, the second great advantage of these programs is the accumulation of customer data, leading to a more complete view of the customer. An example is the Miles & More program, where Lufthansa and its "Star Alliance" partners, hotels, car rentals and others are partnering and sharing their data. All partners can then benefit from the shared usage of this superior data.

When talking about external usage of customer data, privacy issues are a major concern. The legal environment is very complex, differing from country to country. Many aspects are currently under discussion. The US government, for example, strongly objects to the data privacy rules of the European Commission, because their much stricter requirements are said to hurt transatlantic e-commerce. Anyway, in most cases the customers must actively agree upon the intended usage of their personal data. Apart from the complication of satisfying the laws, the possible negative impact on customer relationships must be considered. Yet companies have opportunities to avoid these negative implications by actively addressing the problem. The main claim asserted by data privacy organisations is therefore transparency for the customer. When customers feel well-informed and can rest assured that their concerns are taken seriously, they may well be willing to agree to their data being used. The 7.7 million subscribers of the payback program in Germany impressively demonstrate the opportunity to get the customer's approval.

Excursion: CRM systems

The technical realisation of a customer relationship management system is illustrated in Chart 28, where the whole system is divided into four parts reflecting the system's four main functionalities:

1. The data sources represent the various internal and external data acquisition capabilities.

2. The channel application architecture serves the interactive usages of the system, supporting customer contacts in the sales and service departments. Individual customer data is accessible via CRM application services by all interaction channels and can be updated in real time. These real-time capabilities and the availability of enough information to drive interactive sales and services are extremely important. Furthermore, the interactive functions are usually mission-critical and therefore require high availability of the system and the ability to support high transaction volumes.

3. The marketing application architecture allows analytical usage of customer data, used by marketing and management in order to develop and evaluate customer strategies. Here, access to more complex information is important for driving modelling, general analysis and campaign management. The number of users and queries is typically much smaller than on the customer interaction side. A particular challenge is presented by the complexity of queries supported by the system.

4. The customer data architecture is the heart of the CRM system. It is designed to integrate, store and process all customer data in a way that supports either the interactive as well as the analytical usages of data. At first, input data flows through a data refinery stage where basic edits and validations are performed in order to achieve a consistent data basis (e.g. match-key functionality, de-duplication services, household consolidation, integration of different data formats). The interaction system holds data organised

---

for optimum support for the interactive usages described above. The Data Warehouse, in connection with Extract/Transform/Load (ETL) capabilities, provides data for further analyses with the analytical system, allowing more complex data processing serving marketing and management applications. Finally, the meta data repository promotes integrated performance support (e.g. user help) and reconciles conflicting data definitions.

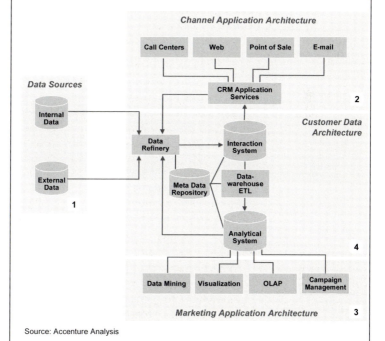

Chart 28: The CRM Solution Architecture

This brief description of a typical CRM system is certainly not exhaustive, but reflects how complex the system needs to be in order to allow the diverse important business capabilities. The detailed structure of individual systems depends on the company's customer strategy. Today, software vendors such as Siebel Systems, Nortel/Clarify or Peoplesoft/Vantive have specialised on customer relationship management systems, offering complex and tailored business solutions. These CRM solutions cover the requirements to a certain extent especially in the customer interaction part. To enable a company to transform data into information by applying certain techniques and analyses, it is vital to consider experienced vendors in the area of analytical CRM, e.g. SAS, IBM, NCR, Oracle, Prime Response, Data Distilleries or Seisint. Most of the tool sets offered by these vendors take into account the special requirements of analytical CRM: scalability, availability and performance.

Main Customer Insight Revenue Streams
New revenues can be raised by selling customer data: customer data can be sold to advertisers and other companies interested in relevant customer data. The market volume for customer data is expected to grow significantly under the influence of the available high-quality data.
Customer data can be sold to partnering companies.
Additional revenues can be raised by using customer insight to enlarge the customer base. Thereby, all revenues paid by the end customers as well as other revenues related to the size of the audience (e. g. advertising) increase.
Other additional revenues come up when customer insight is used to increase each customer's willingness to pay by means of more attractive products and services.
Revenues by transactions as present in all markets, explained in detail in chapter 5.1.3.

> Permission advertising will complement interruptive ads. It will be targeted along customer habits and tailored to interactive content. Due to interactive broadband, new kinds of advertising and therewith new revenue sources become possible.

4.2 Advertising

Get Audience			Sell Audience		
Content Production	Content Aggregation	Content Distribution	Customer Insight	**Advertising**	Merchandizing

The second market in the Sell Audience area is advertising. In the classical definition, advertising is the methodical attempt to win somebody or a specific group for something. It is a marketing instrument used to increase company sales. Advertising instruments in the traditional sense include print, picture, personal mediums, acoustic advertising and shop windows.

There are two generic kinds of advertising: interruptive and permission advertising. Whereas in interruptive advertising customers are neither asked nor voluntarily choose to see ads, but are "detracted" from the things they are actually doing at that time, permission advertising allows customers to give their permission to be shown certain kinds of spots. Classical examples of interruptive advertising are today's TV advertising between movies, newspaper ads before you can flip the page and continue finishing a story or a banner on an Internet site offering you something for free. Permission ads are especially designed for the individual end users; valid examples are advertising related to customer loyalty programs and special personal sale initiatives - all kinds of activities where end customers expressly request a certain kind of advertising.

In earlier years, advertising was mainly permission advertising - people entered stores to be informed about different products, brands and functions - because there was no mechanism to effectively launch mass ads. Only in the later years with the establishment of mass media was the basis for interruptive ads and marketing launched. Advertising and marketing became more interruptive, with permission advertising becoming a rarity.

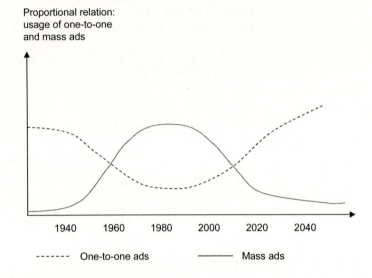

Proportional relation:
usage of one-to-one
and mass ads

1940   1960   1980   2000   2020   2040

- - - - - -   One-to-one ads   ———— Mass ads

Source: Accenture Analysis, Permission Marketing

Chart 29: Development of Advertising

As demonstrated in Chart 29 above, broadband is reversing this evolution. Due to the personalised programming and the possibility for the aggregator to retrieve customer insight, interactive broadband is building a platform for permission advertising again. This platform is constantly gaining importance in the advertising realm and will be used more and more in the near future. Personalised advertising is much more efficient and promising than traditional interruption advertising due to the fact that it can be personalised and creates customer insight. This advantage is not only seen by the advertisers but by the customers to the same extent. The necessity to move forward to new forms of advertising on the Internet is illustrated by Chart 30, in which the declining click-through rates of Internet banners are depicted. Although the development shown may be frustrating, Internet advertising is today an important part of the marketing mix, allowing companies to address a large number of potential customers. Advertisers should not do without it, but rather go for increased interactivity and permission. A successful example is the online travel agency Travel24, which managed to push click-through rates of their online ads up to 11% by using interactive movie sequences. More generally speaking, it has often been stated that advertising budgets will sooner or later follow the media usage.

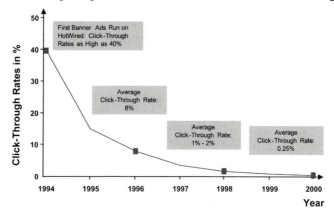

Source: Jupiter Analysis, Internet: Advertising Bureau, Accenture Analysis

Chart 30: Declining Banner Click-Through Rates Demonstrate the Rejection of Mass Advertising

4.2.1   Development of Permission Advertising

Permission advertising is a kind of advertising where the consumer allows a company to send advertising and information. Four factors are vital for the success of permission advertising: personalisation, anticipation, frequency and relevance.

Personalisation is the enabler for permission ads. If customers do not provide their personal data and allow the interactive broadband service to create customer insight, no permission advertising will be possible. Anticipation means that the customers are eager to use the service and to hear from the advertiser. Relevance is the core importance of permission ad-

99

vertising. The customers are willing to see only those kinds of ads and spots they are really interested in. The overall goal of permission advertising in the broadband realm is to create a stable relationship between the advertiser/aggregator and their customers. Therefore, the customer's frequency of usage on the one hand and constant updating of the site on the part of the advertiser on the other hand are essential for a cycle of loyalty and bilateral advantages. This means that frequency is closely linked to loyalty.

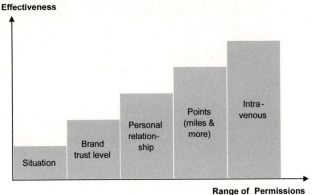

Chart 31: Intensity of Permission

As can be seen in Chart 31, the advertising realm disposes of different steps of permission. Starting on the left-hand side, *situational* permission is given very emotionally and due to a current situation. A valid example is the offering of a related T-shirt or instruction book when a user purchases an interactive video game. This is the less intense type of permission which at the same time delivers the least customer insight and potential for long-term customer relationships.

The *brand trust level* is permission given due to the fact that there is a certain relationship between a consumer and a brand or product. This thinking is very valuable, because it leads to various buying decisions but does not automatically entail deep customer insight yet. This brand affinity has to be exploited in order to bind the customer even more to the brand or product.

A *personal relationship* is the best tool on the way to achieving valid customer relationship management. Customers can be influenced directly by the sales representative or other personal contact. They have a guilty conscience if they use their personal contact or relationship but do not buy anything at the end and, finally, the good personal contact makes it a lot easier to approach the different clients in a very personal way, offering distinct services and products to clients based on their interests and habits. This stage of given permission is already quite intense because customers are opening themselves up towards the company, the products and it's representatives.

A *points* system is the first real customer relationship system represented in this chart. The customers are given a customer card (which can be a traditional paper card for collecting stamps or a modern electronic model), which has to be shown every time the customer uses a special service, makes a purchase at a special store or buys certain products.

The points system can work in different ways. The paper card model, for example, only enables you to track the frequency at which products are bought and the amount of money spent because this is about all a stamp can tell. The best examples of companies that use this "stamp card" are bakeries, coffee shops and similar stores. Customers usually have to submit no more than their name and address to use this type of permission advertising. This means that customer insight is not really retrieved. The electronic-based model goes one step further, requesting more customer information before issuing the customer card. The customer's data is saved on the card and the service is more personalised. The best examples of electronic customer cards are the customer relationship systems from various airlines such as Lufthansa's Miles&More, British Airways' Executive Club, Iberia Plus and customer cards in grocery stores, which enable the company to track every single purchase in terms of products, brand names, date and hour of purchase and therefore create highly detailed customer insight with the aim of providing more personalised services.

The last and most intense type of permission advertising is very appropriately called *intravenous*. It is the form where a customer signs up to receive a special kind of advertising or a special amount of products in a special period of time. The classical examples are "book clubs" and "music clubs" where customers sign up to purchase a number of books or CDs in one year for a certain amount of money. Customers are enticed to sign up by very interesting and luxurious presents. The intensity of permission lies in two different aspects. Firstly, the clubs have guaranteed sales - that means fixed revenues - for a certain period of time and, secondly, they simultaneously have a clearly defined and very loyal customer base with lots of customer insight. People give detailed information when signing up for the first time and constantly update their own data by ordering interest-related books.

The ideal and most efficient method of permission advertising is to lead the customer through the different stages of permission intensity, starting with the situational stage with an initial contact and ending with the intravenous stage, by constantly intensifying the relationship with the customer, personalising the service offerings and maximising the permission given. Today's media variety and especially the interactive broadband services enforce and ease the transportation of the customer from one phase of permission to the next.

An ideal example could be as follows: as a result of advertising on a rainy day, a customer orders a video "on demand", giving away basic personal data. Satisfied with the service the customer orders again, knowing exactly what to expect (brand trust). When trying to order the next time, the VoD portal already suggests different movies to the customer along the lines of "People who watched this movie also liked that movie" (personal relationship). The next step is to retrieve more customer insight and establish a customer relationship model. From this point on, end customers receive points or credits for every movie they demand. In the last step - intravenous - a video is sent to the customer automatically every Sunday.

The Internet was the first medium to allow permission advertising by offering a very low cost and relatively easy means of contacting a huge number of people. The Internet's interactivity enables companies to track their customers' behaviour and interests. Broadband will enforce this trend and bring it into a new dimension.

Knowing what customers buy, like, what they are interested in and how they spend their free time, companies are able to address each customer in a very personal way as does Amazon.com today, for example, by recommending other books with every purchase.

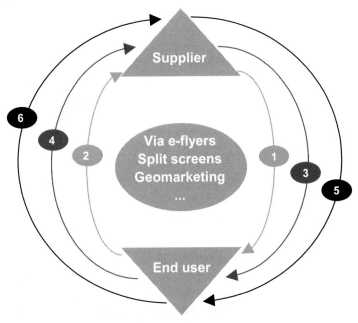

Source: Accenture Analysis

Chart 32: The Permission Advertising Cycle

Permission advertising is leading to a mutual beneficial relationship cycle between the advertiser and the customer. The critical success factor is the establishment of this relationship, i.e. the creation of the initial contact.

Although permission advertising is becoming increasingly important, interruptive advertising remains dominant today. Permission advertising will always rely on interruptive and mass ads to establish the initial contact just described. Since customer insight is a very complex and continuous process, as described in the previous chapter, a relationship has to be created with little or even no knowledge and insight about the customer.

As a result, the first step in the permission advertising cycle illustrated in Chart 32 will always be *interruption advertising (1)* in order to establish the initial contact. The idea is to catch the customer's attention and to generate interest for a specific brand, product, service, etc. Having established this initial contact, the next step is to create *a first sign-up (2)* related to the interruptive advertising, i.e. products & information are ordered, services are used, offers are accepted and requested, people call a hotline, participate in quizzes, etc. Having established this initial contact, the advertiser tries to intensify this relation by committing the end user one step further. *Special offers (3)* are made for every customer, i.e. incentives to reinforce the interest. In return, end users give their *permission (4)*, simultaneously providing the advertiser with their data for personalised information and advertising, which is essential and very useful for the continuity of the permission advertising cycle. Disposing of this customer insight, the advertiser (or aggregator) is able to offer *a personalised service* to customers and *reward* them *(5)* in form of coupons, cash, points, playing games, etc. This personalised service and the rewards program lead to *loyalty (6)* on both sides. The end customers get all the data, information and service that they need and request and therefore have no interest in using any other channel. In addition, they receive rewards, which create additional loyalty to the company. The company thus has very valuable customers, due to the insights they have about them and due to their loyalty. It can be seen that trust is created and a mutual beneficial relationship is established which leads to a circular flow, shown in the graph above by the lines numbered (5) and (6). If this cycle is established, the aggregator/advertiser has managed to "transport" the customer through the different intensive forms of advertising from being a "lower" not so interesting client to becoming a high-potential customer.

The main consequence of this development is that the fastest and most interesting/successful aggregators are used most and build up a market entrance barrier for later competitors. Companies that use interactive broadband advertising will sell more products to less customers, omitting stray and focusing on special and specific target groups.

4.2.2   Advertising Today and Tomorrow

Beside the fact that permission advertising will enter a new era, interactive broadband will lead to innovative forms of interruptive and permission advertising. The new technology enables the bundling of TV spots, commercial-based TV programs, brochures, folders, catalogues, demo tapes, etc. This combination allows new forms of advertising to be created. Examples include:

- *Enhanced and interactive ads:* Interactive movie sequences (superstitials), product information by pushing a button (e-flyer), booking a test drive, making an appointment with a sales agent, follow-up or direct transactions.

- *Geo marketing:* Every set-top box sold has a serial number, which can be used to see where exactly, i.e. in which city, the box was sold. Using this information, ads can be sent regionally, promoting a new store, sale, etc.

- *Advertising in stages:* Based on the serial number concept, advertising can be sent very strategically. Customers who have ordered specific product information or interacted in advertising frequency can be provided with additional information the next time they use the service and will be shown a different spot the next time.

- *Special target groups:* Due to the existence of an increasing number of special-interest programs and channels, advertising will be aimed more directly at its respective target groups.

- *Situational addressing of the customer:* Offering additional, related products or services when a service is used.

- *Internet-like inventory.* Home screens and programming guides feature banner-like spots.

- *In-program sponsorships and interactive product placement.* These elements will allay the increased channel surfing, ad skipping and time shifting.

- *Split screens* with personalised ads

- *Theme ads* within the EPG

- *Brand channel:* also in combination with e-flyers

Looking at the examples described above, it becomes obvious that the way advertising works nowadays is going to be revised totally. Whereas in the past - and still today - advertising has been used to "help the sellers sell", it is going to be revolutionised towards helping "the buyer to buy". Brand channels will start playing an important role as well as new program formats such as advertainment - a mixture of advertising and entertainment.

Advertising in the Interactive World

When interactive advertising was launched in 1996, there were two major platforms on which advertising was based: AOL and the web. As can be seen in Chart 33, this landscape has changed significantly since then. Today, advertisers are facing a broad variety of delivery channels, such as e-mail, wireless and other emerging platforms like interactive broadband.

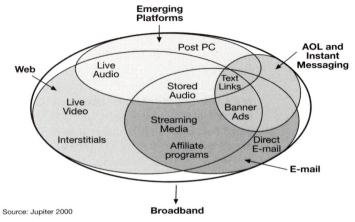

Chart 33: Different Advertising Platforms

Interactive broadband allows an ideal combination of multi-sensual attractive advertising as was explained earlier in this chapter. In the long run, the idea is to have an MCAP that is delivering the same advertisements to whatever device the customer is using at that moment. But before this becomes reality, we will continue to have three different platforms, with different kinds of interactive broadband ad content. The development of these individual platforms is discussed in the following.

1. Advertising on TV

Advertising on TV will start to evolve with the change from analogue to digital TV (this is already done in some European countries). Until this change takes place for the majority of TV households, the only difference from today's advertising will be that ads will promote Internet pages including messages like: "For further information please visit our website www...".

After having completed the transition from analogue to digital TV, the platform for the development from broadcast to narrowcast will be created. idTV will then enable the direct link from traditional TV advertisements to homepages and links. For example, distributors of special interest or brand channels such as hunting, fishing, golf and cooking channels will be able to target their advertising in two different ways. First of all, the existence of these special-interest channels allows direct contact to special-interest groups. As a result, advertising can be very detailed and specialised due to the in-depth knowledge that is assumed of the people using that specific channel on a regular basis. Secondly, the information sent via the back channel enables the distributor and aggregator to track the customer's behaviour and interests, which in a second step allows even more intense targeting to individuals. Based on this knowledge it finally becomes possible to combine traditional TV with other links, e.g. spots or brand channels will be offered on an e-flyer basis. This means that spots will only be shown or the user will only be led to the brand channel if the user clicks on the button for the flyer and therewith gives permission to be shown the "interactive" ad.

An extremely vivid example is the concept of DaimlerChrysler that uses e-flyers to transform the customer from a low level customer into a high potential client. Via the flyer, the company provides customers with product catalogues and guides them to a brand channel that is broadcast on its own frequencies. This channel delivers background information on a specific car model or a movie with a car in action. These channels are pure advertising-oriented programs used to establish client contacts of all kinds.

2. Advertising on the Internet via PC

Advertising on the Internet is set to continue its momentary development, as this line of business is already quite successful today. The company "Doubleclick" for example is providing around 2 million Internet advertisements daily, delivering the issues with detailed feedback and clustered customer insight. This number is expected to grow to about 200 companies and 500 million mails daily. Due to the faster and better transmission of data, advertising will become more complex and developed but will basically stay the same. The entire range - from banners to short advertising movies - will still exist. Only with the establishment of real interactive broadband will ads literally become interactive, enabling customers to select exactly what they want to see and therewith changing the entire realm.

3. Advertising on mobiles:

Nowadays, advertising is being extended to mobile phones and other wireless services. In these cases, ads are sent in the Short Message Service format (SMS) to end customers who have previously given their permission. In return, they receive points that they can convert into premiums, for instance in the case of Deutsche Telekom. Of all those asked, 76% would want to use their mobile for shopping purposes if there were a good relationship between price and value. A Forrester study shows that by 2004 every second mobile user will prefer m-commerce to traditional online shopping. This development will be led by the new UMTS technology. In 2005, when UTMS becomes reality, SMS will be a 'timeworn' format, since it is already possible to send and receive content-rich services and applications as pictures, movies, etc. High-volume broadband information, commerce, entertainment services and advertising will be sent to mobile users.

The recent key word in this context is location-based services. In the context of advertising this means that ads can be sent directly to a handheld device, for example providing product information, when and where it is really needed. Imagining that the handheld device will also be equipped with a Global Positioning System (GPS), products and services will be advertised locally even when the customer is on holiday, on a business trip or at home. These location-based services can be offered in a passive and active way. Passive means that advertising is automatically sent to the consumer, while the active method involves the consumer becoming active and requesting the advertising. A good example of this is somebody driving on a motorway and requesting directions to the nearest McDonald's restaurant or fuel station.

Value Added by MCAPs

Interactive TV is a huge step towards a new advertising realm. This step will achieve even greater momentum in the near future, when the first MCAPs are created and begin to offer initial services. For advertising, this means that people can access their personalised portal not only via their idTV but also via their mobile device or PC.

Via these MCAPs, advertising related to current activities on the screen or promotion information of general interest is provided to the customer in any format. In this case it does not matter whether the user is watching TV or using the Internet. The functionalities of the MCAP allow you to send all kinds of advertising to any device, automatically changing formats, programs and picture resolutions. These opportunities are creating a new market for integrated campaigns. In the long run, this evolution will lower advertising costs due to the limited number of advertising formats used and will increase the effectiveness of advertisements.

4.2.3 **Market Size, Trends and Development of Different Media & Broadband**

Predicting and estimating potential interactive broadband advertising revenue numbers for the coming years in Europe is a complex task, since interactive broadband advertising does not fully exist yet. Hence, numbers have to be affiliated by combining the actual online advertising numbers with actual TV advertising numbers and by anticipating further developments. Based on this approach, it is anticipated that total advertising revenues will grow during the next few years, with idTV overtaking revenues from traditional TV advertising and the Internet. It is not expected to generate new revenues in the short term due to the slow-down and intensified competition in the online and TV advertising arena. Also, the growth is expected to be smooth, especially in the first few years, because of the rollout times of idTV platforms. Referring to a new US-based analysis performed by Jupiter Media Metrix, advertising in idTV is expected to have only a 7 per cent share (estimated at DM 10.2 billion) of the whole advertising money spent in the US up to 2005.

At this point in time it is interesting to see how producers, aggregators and distributors - the Get Audience side - can benefit from these revenues and how they can make use of advertising. The producer can include the different kinds of advertising in productions, for instance, in form of product placement and therewith receive direct revenues. There is also the possibility of producing brand channel content and/or advertising movies together with companies. A good example is BMW, offering short movie commercials at bmwfilms.com. Aggregators should be the ones to benefit most from advertising revenues. They can sell advertising space and contacts on their sites as well as link the user to special ads or sites and participate in sponsoring activities, earning high advertising revenues in all cases. But, as mentioned earlier, a basic product or service is needed. Here, a good example is RTL, which had an advertising market share of 18.2% for the first half of the year 2001 (which constitutes an increase compared with the 16.9% in the year 2000, whereas Prosieben only showed an increase from 13.5% to 14.1% in 2001). Referring to the FTD (19.06.01), the main reason is seen to be successful products like "Wer wird Millionär?". Distributors, as the last link in the Get Audience chain, only have limited chances of participating in advertising revenues due to the fact that they only transmit aggregated content.

Knowing the forecasted market size, it is interesting to give an impression of how online and offline spending is distributed among the different industrial segments. It becomes obvious that Media and Financial services are the two areas that mostly drive online shopping revenues. Therefore, interactive broadband advertising in those areas promises to be excep-tionally effective, whereas consumer packaged goods do not seem to have the best "online shopping" prospects.

The broadband evolution supports one more significant development in the media realm: brand building. The brand building opportunities in today's Internet are quite limited compared to the possibilities interactive broadband is set to offer. The faster access will allow advertisers to replicate the television commercial experience on the Internet. Some of the heavy advertisers in traditional media are expected to shift a large proportion of their marketing budget to interactive broadband. As a result and due to its effectiveness, it is expected that the proportion of online advertising spending will increase during the next few years.

Main Advertising Revenue Streams
Higher advertising revenues due to sophisticated focus of targeting.
More viewers and users due to a highly personalized kind of advertising that the end-user accepts and is willing to see. The result is less "zapping" and higher revenues due to a large client base.
Improved content offerings related to advertising returns. Shift from get to sell. Customers can use a service without paying for it, because the advertiser pays for the service.
Through product placement and sponsoring, portals do not buy content themselves. They feature content presented by other companies. Revenues in those cases are generated from sponsor's payments for the "permission" to show a movie and higher viewer numbers.
Good and personalised advertising leads to higher shopping and transaction revenues.
Revenues by transactions as present in all markets, explained in detail in chapter 5.1.3.

> Targeted merchandising will focus on merchandised products, recommendation and co-branding, along with the brands established around portals and content. Event/intention-centred merchandising will be the key to exploiting additional revenues.

4.3 Merchandising

Get Audience				Sell Audience	
Content Production	Content Aggregation	Content Distribution	Customer Insight	Advertising	**Merchandizing**

"Nothing sells like merchandised products" - this sentence expresses the sales promotion effect of merchandising - the third market on the Sell Audience side of the framework. On the one hand, merchandising is gaining importance as an innovative instrument in the modern marketing mix of manufacturers and retailers. On the other hand, it represents an increasingly relevant revenue stream for content producers and holders of content rights to refinance their content. Thus, it plays a significant role in the process of revenue shifting from the 'Get Audience' to the 'Sell Audience'. As is the case for the other markets on the Sell Audience side, one main prerequisite for generating revenues is at least having a good product or service as basis.

4.3.1 Merchandising in the Media Environment

'Merchandising' in the marketing world is what Anglo-American marketers describe as 'licensing'. In this study, we employ the term according to a more operational usage; this means that merchandising contains sales-support measures occurring through the commercial exploitation of brand popularity. It includes merchandised products, recommendation and co-branding. In order to exploit the brand's potential effectively and

113

to achieve optimal access to customers, the licensees'
merchandising plans have to take into account the different
participants and their target groups involved in the
merchandising process, as outlined in Chart 34.

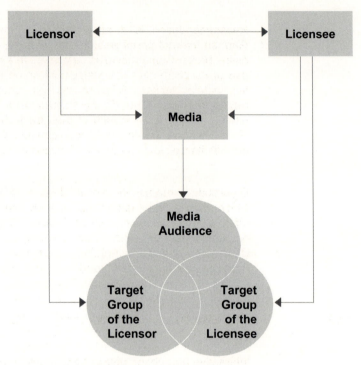

Chart 34: Participants in the Merchandising Process

In contrast to advertising, product merchandising is founded on
a license agreement between the licensor/rights holder and the
licensee/manufacturer. The licensees buy the right to integrate
the licensing property into their marketing and sales activities as
agreed in the licensing contract. Product merchandising aims to
establish an emotional connection between two non-related
items: a brand image (e.g. Big Brother or Bayern München) and
a merchandised product (e.g. a pencil, toothbrush or cap) in
order to strengthen preference and purchase intention. The

transferred image constitutes the product's unique selling position (USP) that influences especially children, teenagers and young families in their buying decision. Merchandised products in the media business therefore often target this key customer segment of the 14 to 49-year-olds.

Recommendation is a special form of merchandising. The German Internet portal web.de - for instance - recommends the online broker Comdirect to its customers in a newsletter, making use of its customer relationship and brand name to advertise the online broker. RTL serves as another example of recommendation. This TV aggregator used its existing brand name to support the launch of the new shopping TV program "RTL Shop" in March 2001, recommending the new program through its well-known name and customer relationship.

Co-branding represents the third form of merchandising, in which two or more established brands are combined in one offer, e.g. the VW Golf edition "Rolling Stones". Each of the brands hopes to reach new customers, as illustrated in Chart 34, and to strengthen their own brand by being associated with the other brand.

4.3.2 The Importance of Merchandising in Interactive Broadband Services

Interactive broadband gives rise to a new form of content (i.e. a new art of storytelling) and service delivery (i.e. real interactive applications), offering new opportunities for merchandising activities, as illustrated in Chart 35.

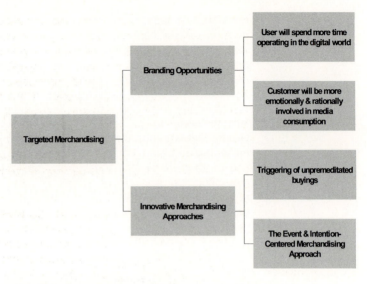

Chart 35: New Merchandising Opportunities in Interactive Broadband Services

Firstly, targeted merchandising in the interactive broadband environment increases the branding opportunities for media products. Two trends are responsible for an enhanced presence and the importance of the brands in the interactive broadband realm. On the one hand, the audience will have a greater opportunity to interact with the broadband applications. Also, the new kind of storytelling will make it more emotionally and rationally involved in the process of media consumption than it is nowadays without interactive broadband access. On the other hand, a broad variety of additional services will be offered via interactive broadband access technology, e.g. idTV shopping, banking and chatting. Using these interactive services, the consumer will spend a lot more time operating in the digital world.

Secondly, interactive broadband allows the realisation of innovative merchandising approaches, triggering unpremeditated purchases and exploiting the new quality of pre-clustered communities. Detailing the first approach, broadband content, like interactive movies or live coverage of soccer games, offers the possibility to initiate impulsive purchases of merchandised products. The situations described in the following serve as examples to elucidate the given statement. While watching a soccer game on idTV, enthusiastic soccer fans decide to buy the tricot of the brilliant goal scorer. They therefore pick up their remote control, click on a few buttons and fulfil their wish via idTV, before rethinking their purchase decision. Other viewers, watching an interactive daily soap, want to buy an umbrella used in the film. They therefore stop the soap, use their remote control to click through an interactive catalogue and purchase the umbrella from the soap's merchandise collection. In both cases, the broadband content serves as a recommendation for the merchandised products, whereby the convenient and easy-to-use consumption process is made possible by the new interactive broadband services. Another very promising merchandising approach involves commercial pre-clustered communities, described in the chapter below.

All things considered, interactive broadband brands - whether they are branded aggregators (e.g. RTL or Focus Online), branded characters (e.g. Mickey Mouse), branded personalities (e.g. James Bond), branded formats (e.g. Big Brother) or other forms of branded media products - will enjoy a stronger brand/name recognition and brand preference than their equivalents in the media business today. They will have a higher potential for merchandising activities, since a brand is the main prerequisite for successful merchandising activity.

4.3.3 Pre-Clustered Communities: The Martha Stewart Example

The company Martha Stewart serves as an excellent example to illustrate the merchandising phenomenon as it exists today. Moreover, it demonstrates which merchandising activities might be possible for content producers, aggregators and even distributors to generate additional revenues in the Sell Audience realm.

Martha Stewart Living Omnimedia produces "how-to" information and content covering all topics related to the home including the eight core content areas: cooking & entertaining, gardening, home, crafts, holidays, baby, keeping and weddings. Therefore, the brand stands for "home and living", and especially serves the interests and needs of homemakers.

The idea goes back to a woman called Martha Stewart. Over time, she was able to establish her name as a brand standing for information and advice for all aspects of the house and home. Now, as an international multimedia company, Martha Stewart leverages the well-known "Martha Stewart" brand name across a broad range of media including magazines (e.g. Martha Stewart Living and Martha Stewart Wedding), books (e.g. "The Best of Martha Stewart Living"), newspaper columns (e.g. askMartha, a New York Times Syndicate column), radio shows (e.g. askMartha, a 90-second daily radio feature) and television programs (e.g. Martha Stewart Living syndicated series and weekly "CBS This Morning" appearances).

The company designs products (e.g. Martha Stewart Everyday Baby line and Martha Stewart Everyday Housewares) that are manufactured by strategic partners and sold through a variety of "bricks and mortar" stores (e.g. Sears and Canadian Tire). The Internet segment of the company consists of a website (marthastewart.com) delivering content - supplementing and supporting the offerings via discussion forums, news, weekly question and answer hours etc. Moreover, Martha Stewart operates in the shopping business. The company runs Martha By Mail, a mail-order and online-shopping catalogue offering and all kinds of merchandise for the typical housekeeper.

Martha Stewart uses all forms of merchandising to effectively exploit the pre-clustered community. Merchandised products carrying the Martha Stewart label are available in approx. 5,000 retail stores in the US. The company uses its media presence to create demand and recommend products. Last but not least, it makes use of the co-branding approach by designing products (e.g. Martha Stewart Everyday Home products, Martha Stewart Everyday Garden products), which are sold exclusively in Kmart stores in the US and Zellers stores in Canada.

The following chart illustrates the different forms of media, direct and electronic business that are arranged around the core content of the Martha Stewart brand.

Source: Martha Steward Living Omnimedia LLC., Accenture Analysis

Chart 36: Martha Stewart as a Pre-Clustered Community

This example demonstrates how the potential of a particular community is exploited to maximise transaction-based revenues. Martha Stewart's revenues amounted to approx. $260 million in 2000, amounting to an annual five-year growth rate of nearly 30 per cent.

4.3.4 The Event- and Intention-Centred Merchandising Approach

Events and intentions provide a communications platform for a specific customer base. The audience at a Britney Spears concert, for instance, has a specific interest in everything about the American pop star, while football fans of Bayern München are interested in any information about their favourite Bavarian soccer club.

The communications platform serves as the basis for community interactions. Around branded stars (e.g. Britney Spears) or events (e.g. a soccer game of Bayern München), communities can be built up, as demonstrated in Chart 37. The inner ring of bubbles around the event/intention displays the different forms of media delivering the specific content to the community. The outer circle of bubbles covers examples of services and merchandised offerings specially designed for the targeted community. By providing such a full-coverage offering, communities can effectively be turned into cash.

Source: Accenture Analysis

Chart 37: The Event/Intention-Centred Merchandising Model

In the context of the event/intention-centred merchandising, interactive broadband content and services provide additional features and sources for closer communication and interaction with the targeted customer base. The new quality of interaction and emotional involvement as described above revalues pre-clustered communities. New forms of content delivery (VoD, PPV, interactive and adaptive films) and personalised service offerings (e.g. special-interest video conferences, individualised one-to-one communication, community chats, games and gambling) broaden the range of offerings for the targeted community. Thus, the purchasing potential of the community can be exploited more effectively with interactive broadband offerings. To sum it up, interactive broadband services provide a new quality of content, Get Audience branding and new targeted merchandising approaches.

In order to employ the new merchandising and branding opportunities, content producers, aggregators and even distributors have to take into account the following key success factors for effective event/intention-centred merchandising:

- a strong brand, product, or service, which secures customer loyalty

- a unique look & feel, which serves as the USP and targets a specific intention or interest of a pre-clustered community

- adequate merchandised products, partners and co-brands, which match the brand's image, look & feel and the needs of the targeted community, i.e. the overall merchandising management concept

Content producers deliver events and intentions (e.g. Big Brother, Wer wird Millionär?, Bayern München), which provide the basis for an intention-centred communications platform. TV producers like Brainpool and Endemol generate approx. 10% of their revenues from merchandising activities, while the soccer clubs Borussia Dortmund and Bayern München generate approx. 15 to 20% turnover from merchandising respectively. This reveals the fact that a strong brand is a prerequisite for successful merchandising: Bayern München is the most popular soccer club, i.e. brand, in the German soccer community, and therefore generates the highest merchandising revenues. This is not an easy endeavour, as is shown by the recent decision of the German media company Kinowelt to restructure its business by selling the problematic business of merchandising and concentrating only on the core business (FTD, 19.06.01). However, the event/intention-centred merchandising approach represents a profitable and promising source of turnover for the content producer/licensor. Bearing in mind the key success factors, content producers will generate additional revenues especially from product merchandising activities in the interactive broadband context.

Content aggregators (e.g. RTL, Yahoo! and FAZ) structure information and services due to the needs of their target audience. Since interactive broadband offers vast branding opportunities for aggregators, as mentioned above, they are in a good position to increase revenues from merchandising. Content aggregators have the chance to leverage their brand names across evolving broadband content and services, according to the principle of the intention-centred merchandising approach. This becomes even more important with the advent of MCAPs. The launch of the new RTL Shop serves as an indicator for the overall trend. Due to their unique position as an aggregator, the dominating forms of merchandising will be recommendation and co-branding.

Content distributors have a fairly weak position for merchandising activity, since most customers do not care how the content reaches their home devices. One option is to team up with aggregators or even become aggregators to make profits in the field of merchandising. Another possibility - as shown in the Martha Stewart example - is to establish a brand of their own (which is not an easy task) in a field specific to their business, e.g. facility management, and then start to build a community around this brand and service.

4.3.5 Market Trends

The market of merchandised products in Austria, Germany and Switzerland amounted to $7.8 billion in 2000. As demonstrated in Chart 38, the sales have grown constantly at a rate of more than 20% p.a. in the last decade. The market has reached a high level and is expected to grow at an annual rate of 10% in the coming years.

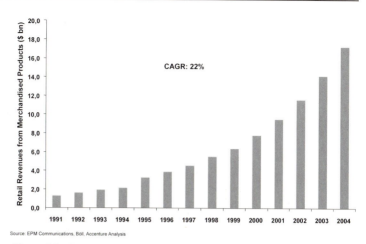

Chart 38: Retail Sales of Merchandised Products in Austria, Germany and Switzerland

Interactive broadband content producers, aggregators and distributors have the opportunity to contribute to the market growth with new revenues, provided they manage to establish well-known brands, successfully target a specific community with their look & feel, offer suitable merchandised products and/or team up with adequate partners and co-brands. In addition to these considerations concerning the profile and offerings, their contribution depends heavily on the mass-market penetration of interactive broadband access technologies and on general customer acceptance/adoption of new merchandising approaches.

Main Merchandising Revenue Streams
Licensing fees are collected for the right to use a brand for merchandised products. The expected sales growth for merchandised products in ASG underlines the statement that there is a huge market potential for merchandised products, and therefore additional opportunities for the licensee to turn licensing rights into cash.
Revenues generated by recommendation and co-branding, often based on a revenue sharing model.
Revenues by transactions as present in all markets, explained in detail in chapter 5.1.3.

5 Revenue Streams and Future Business Models

Having analysed all markets in the interactive broadband services business, we have gained an understanding of this industry. This understanding is crucial for answering the important guiding question as to what a successful business model in the interactive broadband business will look like. In order to present an answer, this chapter combines the industry insights, the intra-market revenue streams described above and other important business aspects in the big picture of business models.

One of the key assumptions of this study is the fact that interactive broadband services are seen as part of all media business. Therefore, the "media business mechanics" have to be applied. The analysis of the Get Audience side showed that huge investments are currently made to be part of the game. But, the current situation in which most content on the Internet and TV is offered free of charge (especially in Austria, Switzerland, Germany) has influenced end customer attitudes and behaviour. As a result, customers are currently not willing to pay sufficiently for access and the content they request, although some efforts are in the process of changing customers' minds (e.g. the German football Bundesliga shown live only on pay TV, paid music downloads and special news on the Internet only for money). In this context, we are already witnessing the growing importance of alternative revenue channels - some are on the Sell Audience side in the interactive broadband arena. Growing importance in this context means that companies increasingly start to subsidise their content offerings in the Get Audience realm in order to obtain much-needed revenues. In addition to this, great opportunities for companies to generate revenues are also arising out of the 'transaction' field. One example is the traditional media company ProSiebenSat.1 Media AG that is now participating in teleshopping activities like Home Shopping Europe (which generated revenues of DM 470 million in 2000 through teleshopping, mainly in Germany), thereby using the new revenue stream of shopping transactions. This issue will be discussed in more detail in section 5.1.3.

Overall, our projection is that only highly specialised companies will survive when exclusively operating in their original core business without getting additional revenues out of other markets in the media business mechanics. Some content producers are in a particularly good position to follow this strategy, whereas content aggregators or distributors are finding it rather more difficult to do this. Companies relying on so-called focused business models will be successful only under certain circumstances, for example strong market leadership, high value proposition, excellent brand or other individual success factors. In contrast, many other players are trying to integrate intelligent combinations of revenue sources from all markets of the media business mechanics. AOL Time Warner is a prominent example. Most companies are said to be better off with this integrated type of business model. Currently, the industry trend is towards further integration.

In order to lay the foundations for a successful design of future business models, we will first sum up previously identified, existing and new revenue streams in the Get Audience and Sell Audience markets, supplemented by the transaction revenue opportunities present in all markets. As we will focus on the revenue streams currently under discussion, the list will not be exhaustive. Following this, the discussion will shift to the business model issue and the specific aspects of the interactive broadband service environment. Then, implications for and types of future business models in the interactive broadband services industry will be presented. Finally, this section closes by looking ahead to the development of the whole industry: Quo vadis interactive broadband media?

Revenue Streams

5.1.1   The Get Audience Side

Chart 39 provides an overview of who participates in which of the main revenue streams in the Get Audience field. The main revenue streams are listed in the first column; the specific opportunities for content producers, content aggregators and content distributors to source these revenues are evaluated in the following three columns. In the first line, for example, it is indicated that sales of content to aggregators are seen as a substantial revenue source for content producers (in fact this is their core business). Content aggregators can generate these revenues to some extent, too, by selling content to other aggregators, whereas content distributors will probably not be able to make money in this way, because in most cases they are not likely to own content. However, the total amount of the individual revenues is not indicated in Chart 39, only the distribution among the market players.

Main Revenue Stream	Opportunity for... (○ = weak, ● = huge)		
	C. Producer	C. Aggregator	C. Distributor
Sales of content to aggregators	●	◑	○
Sales of content to distributors	◑	◕	○
Sales of special content for business applications (e-learning, business TV,...)	◕	◕	◑
On-demand content sales to end-customers (PpV)	◑	●	◕
Video on demand	◕	●	◕
Subscription fees of end-customers	◕	●	●
Licensing: Sales of rights to other content producers or aggregators (especially foreign countries)	◕	◕	○
Syndication: Sales of copyrighted content to other editorials or aggregators	●	◕	○
Service fees for leading traffic	○	◕	◑
Service fees for CRM/billing/maintenance	○	◑	◕
Games	◕	●	○
Gambling	◕	●	○

Source: Accenture Analysis

Chart 39: Main Get Audience Revenue Streams

Based on our experience, the importance of the revenue streams mentioned differs in terms of their revenue contribution. Video on demand, for example, is one of the most popular services but is not necessarily expected to be a main source of revenues (this will be different for hardcore erotic videos). In contrast, sales of content including licensing, syndication and games & gambling are expected to be the biggest revenue sources on the Get Audience side. In the B2B area, sales of specific content (such as e-learning) has the potential to be a big revenue source, too.

Looking at the revenue distribution and the relative importance of the individual revenue streams, it shows that content aggregators can participate in all Get Audience revenue streams and are in a strong position. Content producers are in a strong position, as well. They stay at the beginning of the chain and are able to follow a strategy concentrating on content sales, licensing and syndication only. From this Get Audience point of view, content distributors are facing a problem: they participate in only a few of the Get Audience revenue streams. With the exception of the B2B part, these include mainly subscription and service fees, which are not predicted to be major sources in the long run.

This may have implications for all Get Audience players in general and the content distributors, due to the mentioned difficulties, in particular: the importance of thinking about strategic alternatives. We will discuss this issue in further detail after having analysed the other possible revenue streams.

5.1.2   **The Sell Audience Side**

As can be seen in Chart 40, it is found that content aggregators seem to be in the strongest position with solid revenue potentials in all Sell Audience markets, whereas content distributors are facing the least opportunities. Content producers are able to participate in revenue sources depending on the strategic positioning and the contracts with aggregators. This judgement holds true when the relative contribution of these revenue streams is considered: advertising is clearly an important factor, with its main part coming from selling advertising space. Other important revenues are expected to be generated in the merchandising area, where substantial growth potential is predicted. Customer insight is to be seen mainly as an enabler for all kinds of targeting and as an alternative currency for other services, whereas its direct revenue contribution is limited to some niche players.

Market	Main Revenue Stream	Opportunity for... (○ = weak, ● = huge)		
		C. Producer	C. Aggregator	C. Distributor
Customer Insight	Customer Insight data sales (externally to advertiser, merchandiser, ...)	◑	●	◑
	Customer Insight data sales to partners	◑	●	◑
	Additional revenues due to enlarged customer base	●	●	◐
	Additional revenues by increasing end-customers' willingness to pay	●	●	◐
Advertising	Advertising space sales (show advertisment)	◕	●	◑
	Advertising contact sales (click-through)	◐	●	◑
	Advertising production	●	○	○
	Sponsoring	◕	●	◑
	Product placement	◕	◑	○
Merchandising	Revenues for licensing merchandised products	●	◑	◑
	Revenues for recommending products	◐	●	◑
	Revenues for co-branding products	●	●	◑

Source: Accenture Analysis

Chart 40: Main Sell Audience Revenue Streams

RTL Newmedia is an example of how to successfully integrate several Sell Audience revenues. The company operates the entertainment portal RTL world, which has become the most frequently visited German website. RTL Newmedia is a 100% subsidiary of RTL Television, coordinating all interactive and digital activities. Out of RTL Newmedia's total revenues of €46 million in 2000, 40% were generated by merchandising activities alone. Here, T-shirts, caps and other merchandised products bearing the strong and popular brand "Big Brother" contributed the most.

5.1.3   Transactions

Market	Main Revenue Stream	Opportunity for... (○ = weak, ● = huge)		
		C. Producer	C. Aggregator	C. Distributor
T-Commerce	Interactive shopping (lean forward)	◐	◑*	◕*
	Impulse buying	◔	●*	◕*
	Banking and brokerage	◑	◑*	◕*
	Teleshopping (lean backward, e. g. H.S.E)	●	●*	◑*
	Customer hotline	◐	◐	◑
Paid Information	Location-based services (guiding, tracking, emergency, finding, ...)	◑	●	◑
	Micro paid information (horoscopes, weather, ...)	◑	●	○
Communication	Chat	○	◔	◐
	E-mail	○	◔	◐
	SMS	○	◔	◐
	Voice over IP	○	○	●
	Internet services	○	◑	●
	Videoconferencing	○	◑	●
	High-bandwidth connection for business applications (LAN)	○	○	●

Source: Accenture Analysis

*Revenues paid by producers/aggregators, based on revenue sharing or service fees models

Chart 41: Main Transactions Revenue Streams

Coming to transaction revenues (see Chart 41), three basic activities can be distinguished: T-commerce, paid information and communication. T-commerce consists of all shopping transactions enabled by interactive broadband services including telephone hotlines and banking. Paid information includes various location-based services and micro-paid information, as for instance the latest soccer results or weather reports. Communication, like telephony (voice over IP), SMS, high-bandwidth connections for business use or videoconferencing, can generate additional revenues especially for content distributors leveraging their infrastructure.

When evaluating the relative size of the revenue streams, the first observation is that in the short term distributors are said to generate fairly large revenues from communication. Yet this is not expected to continue endlessly: because of inflating broadband capacities, due to the different access technologies competing in this market, it is likely that this field will become a commodity and prices will slow down. In the long run, it is anticipated that most revenue contribution will be achieved through several forms of shopping and through location-based services. Whereas shopping revenues are already an important part of many business models and still gaining in importance, the full potential of location-based services is still an upcoming topic. As can be seen in Chart 41, these particularly promising revenue streams are mainly within the reach of content producers and aggregators. Producers and aggregators are therefore predestined to add these interesting revenues to their businesses. To some extent, however, distributors also have the opportunity to participate in these revenues at least via revenue sharing or service fee models with producers and/or aggregators as their partners.

As an example of integrating transaction revenues, RTL Newmedia generates 31% of its revenues in the area of hotlines and teletext. In addition, the RTL Group launched the RTL shop, a teleshopping channel with additional service offerings such as banking. The relatively high investments of about DM 65 million are expected to break even within three years. Another example is the channel tm3 aiming to achieve a revenue contribution of 30% by means of teleshopping and expensive hotlines that customers call for shows, online betting and public-opinion polls.

After having reviewed the revenue opportunities in the three main areas Get Audience, Sell Audience and transactions, it can be seen that the aggregator business is expected to be a good place to be in the short run. The projected revenue sources in the different areas show attractive opportunities. In the long run it becomes a brand play where only the strong ones will survive. And, further in the future, time-to-market will determine their success on the way to becoming MCAPs.

Content distributors are in an overall limited position. The market situation leads to an array of issues summarised in Chart 42, leading to the given implications on their business models. Besides these necessary implications, the limited revenue potential in the Get Audience as well as in the Sell Audience realm forces many distributors to extend their operations in the area of transactions, offering mainly telephony and other communications services. If these activities are to become a commodity, three main strategic opportunities for distributors present themselves: (1) they can either focus on their core business and become big enough to realise economies of scale and participate in the volume business or (2) integrate (buy, co-operate) in other markets mentioned in the media business mechanics or (3) enter new markets with new revenue sources.

Issue	Implication for distributor's business model
New competitive market, huge investments	First mover/first follower advantage
Commodity	Compete by value/price or in niches
Prices declining	Offer customer value instead of price war
Disappearing distance/mobility/coverage	Build alliances across regions and technologies
Price sensitivity of customers	Personalisation and package bundling
Transmission quality	Invest in network upgrade or get out of biz!!!

Source: Accenture Analysis

Chart 42: Issues and Business Challenges for Content Distributors

Becoming big enough means, for example, spreading capacities all over Europe or even the entire world. This is one strategy that UPC, NTL, Callahan or Astra are following. Besides this, it is obvious that there is a strong tendency for distributors to extend operations towards further Get Audience activities, mainly in the content aggregation market. Partnering with companies already operating in other Get Audience markets to extend their own activities is an interesting option, too. One example is the recently created UPC Media Division, combining the activities of the distributor chello broadband n. v. with the UPCtv operations. UPC Media Division manages four business units, thereby integrating revenues from all areas: broadband Internet access, interactive services, transactional television services and pay television channels including content production activities. Another example is the possible co-operation between the world's largest media company, AOL Time Warner and NTL, UK's leading distributor. Here, a strategic alliance concerning broadband activities is under discussion. Entering new markets entails expanding business for example into the area of facility management. Here, content distributors are moving in a strong position to compete with other existing players and this opens up new additional revenue sources.

Content producers seem to be in a good position. It is clear that only content that really adds value to the customers can lead to substantial revenues paid by end customers. On the other hand, "content is king" and forms the basis for successfully attracting the audience for aggregators and distributors. Therefore, content producers can sell their content to both (they then can refinance it via Sell Audience revenues) whether they will be paid by the end customer or not.

But content production and management is very expensive - especially high value content (i.e. a one-hour TV drama production costs $1.2 - $1.6 million, while a half-hour TV sitcom production costs $800,000 - $1.2 million) - and it is not certain whether revenues coming exclusively from aggregators and distributors in the broadband arena will cover these costs and allow the companies to stay in business. To some extent this might lead to a concentration process expecting mergers and acquisitions in this area. The problematic situation can be seen in the recent discussion about the cash situation of some companies belonging to the content production market like EM.TV. A general implication can be concluded from this situation: it will be important to use content cross-media to gain in efficiency. The expensive content production is not restricted to being used only once. Large opportunities exist in using the same content in various formats and distributing it to various devices. A lot of the journalistic output of the Axel Springer Verlag, for example, can be used not only in the newspaper Bild but also in other formats like a broadband portal. This idea is realised with the newly formed portal bild.t-online.de, where both companies are partnering to profit from synergies: the content can be sold at least twice, and the portal benefits from offering popular content (and of course from the popular brand). The same effect can be seen when movie productions are re-used as video, DVD or soundtrack, later shown on TV, used for fan-magazines, etc. In addition, the content can be leveraged by the low-cost production of a TV series based on the movie. Another outcome of the cross-media approach is the opportunity to offer multi-channel access, which entails using the original content on TV, PC, mobile phones and PDAs. Here, the content has to be adapted to the appropriate form of storytelling in each case.

5.2 Business Models in the Interactive Broadband Services Environment

Having discussed the diverse existing and new revenue sources in the individual interactive broadband services markets, we finally enter the business models arena. Here, the gained insights will be summarised to adequately address the difficulties and opportunities of doing business in this promising market. With respect to the frequently misleading usage of the term "business model", we include a brief explanation of this concept.

5.2.1 What is a Business Model?

A business model is an organisation's core logic for creating value and making money (see Chart 43). Its main components include a set of *value propositions*, describing how the company plans to create differentiated value. It consists of three parts: the *market offerings*, *non-customer relationships* and *economics*. The value proposition is of central importance to every business. Without a good value proposition, no sustainable success is possible. Businesses based on exceptional marketing only will not succeed in the long run. Also, it contains a set of corresponding *capabilities*, describing the core combinations of people, knowledge, processes, technology and assets used to create the value. Surrounding factors affect the individual shape of every business model. Firstly, business models are embedded in a *business context*. Political, economical, sociological and technological aspects of the organisation's environment are taken into account, as well as its relevant industry structure. In this guide, this is discussed with the media business mechanics in some detail. It will not be addressed any further in this section. Secondly, the *customer decision process* plays a key role. All factors influencing the decision process of customers merit a detailed analysis. Thirdly, the organisation's *strategic intent* greatly influences the shape and the design of the business model. In fact, the business model is often referred to as "the manifestation of strategic intent for an organisation".

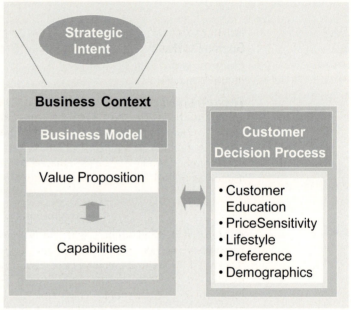

Source: Accenture

Chart 43: Business Models: Main Components and Influence Factors

It is often unclear what makes a business model successful. The success of a business model is determined by its sustainability in the industry competition. Therefore, successful business models:

- offer unique value. This may be a new idea (innovation), or, more often, a combination of product and service features that offers additional value: lower price for the same benefit or more benefit for the same price.

- are hard to imitate. By establishing a key differentiator, such as customer attention, superb execution or a strong brand, these models build barriers to entry that protect profit streams.

- are based on accurate assumptions about customer behaviour.

5.2.2 Business Model Components of an Interactive Broadband Service Provider

Value Proposition

The value proposition offers an explicit description of the unique value that the enterprise provides to its target stakeholders (including customers, partners and stakeholders). The value proposition of the players on the Get Audience side in particular has already been discussed in detail. Only a few short remarks should be added.

With regard to interactive broadband *market offerings*, the main issues are content/service and the kind of content offering. Everybody knows that "content is king". This has been discussed throughout this guide in detail. Here, it should only be mentioned that the content itself has to add value and/or to bring fun or entertainment. If this is the case, people will like the content and will be willing to pay for to a certain extent. Also, it forms the basis for further revenues from the Sell Audience side. The kind of content offering leads to the point how broadband services should be sold to customers. Here, content bundling is the key word. It describes the concept of designing packages of different kinds of content (e.g. high-speed Internet access, telephony and digital TV), which are combined in an intelligent way and sold together. The kind of bundling of the different services/contents will determine the success and the profitability of companies in this market. User friendliness, convenience and cost efficiency are other factors influencing the perceived value of market offerings.

Capabilities

In terms of the capabilities the company uses in order to deliver the proposed value - with regard to the fast growing interactive broadband market - the following additional aspects are of particular importance and should be integrated into a successful business model:

- Because the media industry is on its way to becoming a digital world, Digital Content Services (DCS) is a hot topic. DCSs are focused on the creation, management and delivery of digital content through various devices over any communications channel. Among the hot topics, we see Digital Rights Management (DRM, see page 36), where the proprietary issues of content are addressed, related to the asset *intellectual property*. Content creation, packaging and management is concerned with all the interrelated *processes* necessary both internally (e.g. in order to update the company's own website) and externally, where the workflow between content producers and aggregators, or between aggregators and distributors has to be controlled. A third point addresses the *technological* capabilities, the content delivery networks. Here, the broadband infrastructure, digital TV platforms, data warehouses, aggregation engines and additional functionalities such as search engines and search agents are needed in order to deliver superior products and services to customers and/or to enable effective management and decision-making.

- Thinking about *organisation and people*, it is obvious that diverse skills and personal capabilities are necessary to enable effective internal operations. Other aspects include the relationships between leadership, skills, culture and values, structures, roles and responsibilities, incentives and measures.

- One of the most valuable assets in the interactive broadband environment is a strong *brand*. As in the traditional media business, this idea plays a key role in success. When fully exploited, brands have tangible pay-offs. Valid examples include automated buying processes as can be seen at companies like Coca Cola and Pepsi. Due to a high brand affinity, customers choose the same brand every time they buy a Soft Drink - either Coke or Pepsi. A brand is built over time through a deliberate management process involving strategic decisions and corresponding actions. It distinguishes a company's products and services from those of the competition by combining fundamental factors such as value, customer benefit, positioning, advertising, promotion and a point of sale in a very specific way. Only well-known, branded programs, formats or aggregators can set themselves apart from competitors and attract a broad audience. The brand is the proprietary visual, emotional and rational image that is associated with the media product. For instance aggregators like "RTL" and "ProSieben" bring to mind certain attributes like good entertainment or attractive sports event coverage, whereas "ARD", "ZDF", "Die Zeit" and "FAZ" stand for reliable news and reports. Branding poses several challenges to the marketing strategy to create a high degree of name recognition and strong mental and emotional associations for a new interactive broadband media product. Successful examples in the TV sector are popular formats like "Harald Schmidt Show", "Big Brother", "Wetten das…?" and "Wer wird Millionär?", all promising to deliver a specific set of features, content and entertainment to the audience at all times. An interactive broadband brand does not necessarily have to be a complete new brand, but can also be created by transferring image, trust, etc. from a traditional brand into the new interactive broadband world. Given that time-to-market is an essential factor on the road to success, this transfer may even be more successful than the creation of an entire new brand. The RTL Group, for example, has announced its intention to use its well-known brand name RTL to establish an MCAP named RTL World as an interactive broadband platform.

5.2.3   Customer Decision Process

How will the end customer decide? The expected customer behaviour has a considerable impact on the design of every business model. It gains in importance whenever new markets evolve, and no facts about the customer's decision process with respect to the new products or services are at hand. This is the case in the interactive broadband service market where only limited knowledge and experience about the acceptance of new services and customer behaviour has been collected so far. Nevertheless, predictions of what customers really value and are therefore willing to pay for must be analysed thoroughly. Looking at today's situation in the United Kingdom, where 26% of the population already enjoy access to digital interactive TV, and integrating the results from several consumer surveys in Germany and other research reports, the following five factors illustrated in Chart 44 are important: the customers' demographics, their individual lifestyles, their degree of education concerning the new products or services, their price sensitivity and finally their personal preferences. An individual evaluation of these factors with respect to the new interactive services will determine the sales approach.

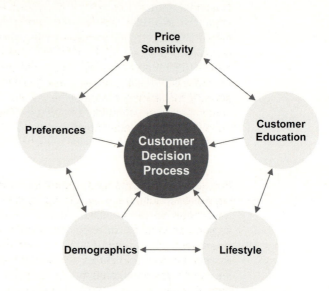

Chart 44: Customer Decision Influencing Factors

Demographics

The main demographic factors influencing the usage of interactive broadband services are often said to be age, income and available time. Concerning age, it can be stated that the age group from 12 to 54 is the most likely to adopt interactive services. This group constitutes the most attractive target group for advertising and merchandising. Interest in news and information is shifted to slightly older people (19 to 54 years), while chat, games and education services are more appealing to younger people aged 6 to 24 years. People older than 54 years are far less interested. The distribution among income groups is positively correlated to income, with high-income groups showing a significantly higher penetration rate than medium-income groups and low-income groups showing the lowest penetration. Anyway, ongoing surveys in the UK show that these demographic differentiators are loosing in importance over time.

Lifestyle

Customer adoption behaviour is considered to be a particularly important aspect of lifestyle. Do the customers belong to the early adopters, enthusiastically buying new products as soon as possible, or will they wait until most of their neighbours are well equipped? Further lifestyle characteristics, as high willingness to take risks versus complete risk avoidance, are related topics strongly influencing customer behaviour. Other factors are the customers' basic attitudes and values. These lead to individual behavioural differences such as the fact that some people read newspapers while others prefer watching television.

For interactive broadband services, many suggestions for segmenting customers according to their individual lifestyles have been presented. Datamonitor, for example, presented a segmentation into 5 groups, according to some basic personal characteristics. These groups are named Rhinos, Gazelles, Pumas, Gorillas and Jackals, whereby these animals represent the identified characteristics. For example the "Rhino" group is described as stubborn, old and set in its ways. Typical members are old, single people. All segments differ in size and, more importantly, in individual interests. As an example, the Gazelles (who move in herds, are timid, do not take the initiative and are typically young families), show the highest interest in games by far. Another promising approach recently published by EMNID concentrates on two separate groups: so-called "couch-pota-toes" versus "PC freaks". The results of their survey revealed the completely different adoption behaviour of these groups. Couch potatoes (characterised by a daily TV usage of more than three hours and the lack of any online affinity) are mainly interested in time-shifted TV, background information concerning the TV program and additional channels, whereas the PC freaks (watching television less than 1 hour per day, but frequently online) value additional services such as high-speed Internet access, e-mail and online banking. This example shows the relevance of individual lifestyles with respect to the customer decision criteria. Other lifestyle-related segmentation approaches focus on customer attitudes towards television: while some customers, for instance, see television as part of their social life, others believe that TV is good only in moderate quantities. Both attitudes have a considerable influence on their adoption behaviour.

Customer Education

This point focuses on the level of customers' previous knowledge regarding the product or service under decision. Are the customers already familiar with the product or service, or do they have relevant experience with similar products or services? When completely new products or services are introduced, understanding customer education often reveals great potential for influencing their decisions.

We have found strong evidence that most people only have marginal previous knowledge about digital TV in general and interactive services in particular. In the UK, for example, there are 12.1 million digital TV subscribers. Only 5.1 million of these are aware of the fact that they have online access via their DTV, and only 0.9 million of these use it at all. In comparison, 10.7 million adults access the Internet using their PCs. Accordingly, many people (34% in UK) said they are not interested in buying digital TV because they "don't need additional channels", obviously unaware of the many other services available within interactive broadband.

Among the German population, previous knowledge seems to be at the same level. According to an EMNID survey, only every fourth German is familiar with the word "interactive television". In contrast, people are widely interested in specific interactive services such as time-shifted TV programs (48% of the German population 14+ years old are "very" interested or "quite" interested) and high-speed always-on Internet access (41%).

In other studies, video-on-demand and pay-per-view applications are often found to be very attractive as well.

Generally, whenever customer demand is evaluated by means of surveys, it is clear that the level of previous knowledge strongly influences the results, because customers can only demand products and services about which they at least have some knowledge. This "bias" effect may have influenced the result of other surveys that named video on demand and games as the most requested applications, in this interpretation partly due to their easy-to-understand and common names. As a result, other, more sophisticated services might be more requested than they currently seem to be if they were better understood.

Price Sensitivity

Customer price sensitivity obviously determines their decisions to some extent. One aspect influencing price sensitivity is whether the offered value is in line with the price, i.e. whether the price is perceived as fair and reasonable (value for money). Another factor is the psychological influence of the actual economic circumstances. When stock prices rise, customer willingness to spend money is likely to increase.

Some hints imply that many customers are quite price sensitive with respect to interactive broadband services. In the UK, 25% of non-subscribers said the high prices for subscription and equipment are their main entrance barriers. For Germany, studies and analyses found the customers to be very price sensitive as well. This reflects the extremely good free-TV supply in Germany, where most customers can access about 30 channels for free. Therefore, the optimum price for interactive broadband services is expected to fall in the range between DM 12 and DM 30.

However, with respect to the discussed education issue, it remains unclear whether customers would change their minds if the real value of interactive broadband services could be considered more adequately, in other words, whether they really cannot afford the price or simply don't see the value for money.

Preferences

Customer preferences take into account personal evaluations of product or service features such as brand, design and the perceived quality. Other general preferences include buying frequency, importance of attributes like trendiness and the desired level of complexity. Whereas some customers, for instance, buy mobile phones with integrated mp3-player and organiser functionality, others prefer simple and easy-to-use models.

Regarding interactive broadband services, customers have individual preferences. The mentioned features especially valued by couch potatoes in contrast to those of the PC freaks are an example. Other preferences vary by country. While US online consumers prefer simple iTV features like pause/rewind, German households said they would like more control over what they watch. Other general preferences are still unclear, e.g. if customers prefer TVs with online functionality to PCs with TV functionality or vice versa.

This leads to aspects of cultural differences as a determining factor for customer preferences. Cultural aspects and differences influence the customer's set of preferences. Looking at content distribution, the business models of, for example, Telewest UK, Telekom Hongkong and the first steps of business development in Germany are completely different, mainly caused by cultural differences.

Combining all of this information about what influences customer decisions, we see two main implications for interactive broadband service business models. Firstly, interactive broadband service providers must consider how to deliver content that really adds value to the variety of customers. Here, different customer preferences require careful segmentation of customers in order to make the right, targeted offerings. This is of double importance, since the old rule saying content is king still holds true: offering the right content is decisive in order to attract the audience. Therefore, this is not only vital in order to receive direct revenues on the Get Audience side, but forms the prerequisite for all other Sell Audience and transaction-based revenues as well.

Secondly, the lack of information about what people can expect from their interactive broadband services shows that educating customers has great potential to increase demand.

5.2.4 Strategic Intent

A business model must always be viewed in relation to the strategic intentions it represents. When a business model is designed, it must be aligned with the strategic goals a company is willing to achieve. This may for instance be the idea of becoming a market leader in one segment, while in other cases it may include the idea of participating in an activity as a niche player. In the language of the media business mechanics, this means that a company may want to stay more or less in one market (i.e. focus on content production as does Kinowelt or be a shopping channel like Home Shopping Europe) or may intend, for example, to operate in the distribution and aggregation market like UPC.

Although strategic intent plays a key role when it comes to developing a business model, it is always to be seen in line with the individual business context. For a successful business model, it is essential to synchronise the strategic planning with the relevant context. In some cases, strategic goals have to be adjusted and changed due to the context a company is facing. As a result, it can be said that business context and strategic intent are the major influencing factors of the business model, even more so since they are strongly linked to one another. This is something that can currently be observed in the interactive broadband market where some companies are realigning their strategies due to changes in business context.

Knowing that every business model is backed up by a certain strategic intent, it becomes obvious that business models tend to differ very much from one another. It is rare to find exactly the same business model with the same strategic intent twice, since that would lead to direct competition with the same products and services. Therefore, it can be said that there is no 'one' universally valid and successful business model but many different ones that will be sustainable in the interactive broadband competition.

5.3 Business Models for Future Interactive Broadband Services

Now, having discussed individual revenue opportunities for and the main components of interactive broadband service companies, it is time to consider the implications for a future successful and sustainable interactive service business model. To begin with, this section will concentrate on the most important general implications, valid for business models for all companies acting in the interactive broadband services environment. The second part will discuss the issue of how to combine the different revenue streams from a strategic viewpoint to form generic types of sustainable business models.

5.3.1 General Implications

If the main findings concerning the most important issues in the interactive broadband services environment discussed in some detail through the whole study become aggregated and condensed, eight general implications for business models in the broadband world appear. These implications neglect other more specific aspects already discussed, e.g. the distributors' need for a strategic decision on how they should compete in a commodity market. The specific implications can be found throughout this study. These eight general implications showing the way to future business models are presented in Chart 45.

NO GET AUDIENCE EXCELLENCE – NO BASIS

Heavy competition for the Audience's attention requires true Get Audience excellence in order to attract a significant Audience as the necessary basis for all further business activities.

NO ADDED VALUE CONTENT – NO MONEY

Content the end customer should pay for must offer real added value, either by quality, exclusiveness, or entertainment. Attractive product bundling is an additional option.

NO CUSTOMER EDUCATION – NO DEMAND

Demand for interactive broadband services is limited by the customers' marginal knowledge concerning these new products and services. Education will increase demand and willingness to pay.

NO CUSTOMER KNOWLEDGE – NO SALES

Only deep customer knowledge allows to appropriately define customer segments and to find a successful way to addressing their preferences with the right product and service offerings.

NO BRAND – NO SELL AUDIENCE

A strong brand is needed in order to enable the full potential of Sell Audience revenues. Creation of this brand is supported by going to market fast or by leveraging an existing brand.

NO CUSTOMER INSIGHT – NO SELL AUDIENCE

Building up customer insight is the necessary prerequisite for generating optimum advertising and merchandizing revenues.

NO SELL AUDIENCE – NO ADDITIONAL REVENUES

The various sell audience opportunities of interactive broadband services must be integrated in order to exploit the full revenue potential.

NO TRANSACTIONS – WASTED REVENUES

New, additional services must be offered to seize transaction revenues, the third wide field of additional revenue opportunities.

Source: Accenture Analysis

Chart 45: General Implications for Interactive Broadband Service Business Models

5.3.2 Five Generic Types of Business Models

Regarding the three main areas for generating revenues - Get Audience, Sell Audience and transaction - five basic types of business models can be found: the Get Audience model, the Sell Audience model, the transaction model, the balanced model and the focused model. While the balanced model aims to integrate revenues from all three areas, the other models more or less emphasise one specific area. Thus, the Get Audience model, the Sell Audience model and the transaction model are currently common. They emphasise one specific area but without exclusively focusing on it. In contrast, the focused model concentrates on one single area (usually on Get Audience markets) and leveraging the core competencies of a company. Before discussing the five types in greater detail, the following points should be asserted:

- The five models are not exhaustive. These are more or less generic types. Other combinations are possible, e.g. businesses based equally on two of the three fields or generating revenues out of new fields that are not part of the media industry, such as facility management.

- There is no "one successful" business model. None of the five is completely right or wrong in general. But some are more consistent (focused model) or exhaustive (balanced model) than others. The appropriateness of individual business models always depends on the strategic intent.

- Business modelling is more than just combining revenue streams, but revenue is the basis for every business model. Hence, we took this as the main indicator for talking about generic types of such models.

- The size of the circle in the illustration has no relevance at all. Only the relative revenue contribution is considered.

- The examples provided represent the named companies in their current state. As interactive broadband services evolve, we expect the revenue compositions of these companies to change.

The Get Audience Model

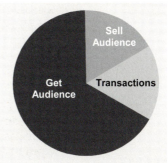

Example : Premiere World

This model emphasises Get Audience revenues, implying that content plays a key role, but tries to earn some money out of other sources, too. The pay-TV model of Premiere World, collecting most revenues via subscription fees and content sales to customers (pay-per-view), is one example. The major advantage of this model is quite obvious: it delivers content free of interruptions, thereby offering additional value to the customers. Here, the content in itself can be presented in a more convenient and convincing way. A second advantage is the opportunity to offer special-interest content, because in some cases, part of the audience is willing to pay high prices for special events (e.g. sports or concerts).

However, depending on the local situation, the revenue potential of this business model seems to be limited due to the limited willingness of customers to pay for content or access, as discussed earlier. In addition, all the kinds of value-added content that customers may be willing to pay for are usually associated with high production costs. What's more, this model suffers from the fact that revenues will shift from the Get Audience to the Sell Audience and the transactions revenue side is not fully exploited.

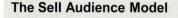

The Sell Audience Model

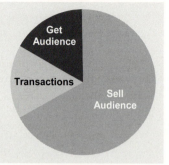

Get
Audience

Transactions

Sell
Audience

Example: RTL Group

The Sell Audience model, with special emphasis on the three markets of customer insight, advertising and merchandising, has been chosen for example by the RTL Group, which generates most revenues in these areas. The main reason for its success can be seen in the fact that it takes into account the limited willingness of customers to pay for content, becoming independent of revenues paid directly by the end customers. In addition, it paves the way to the intention-centred marketing approach, where every event forms the basis for diverse advertising and merchandising activities leading to substantial revenues.

Nevertheless, this strong emphasis on the Sell Audience side leads to a major disadvantage: the dependence on Sell Audience markets. When advertising budgets do not develop as expected, this will have a substantial effect on companies based on the Sell Audience model. In addition, a strong brand is also vital for success. Although this point is true for every media business, the brand becomes even more important when you try to exploit the full potential of Sell Audience revenues.

The Transaction Model

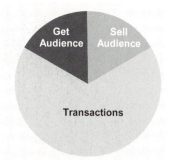

Example: Q.V.C.

A business model with special emphasis on transaction revenues is demonstrated by the shopping channel Q.V.C. Here, the main revenue source is teleshopping. Another strong revenue source in the transaction realm is provided by the expensive telephone hotlines that customers call in order to participate in shows, quizzes, online betting or public-opinion polls. On the one hand, this model has the advantage of low upfront costs. Especially when it comes to shopping, it is a clear advantage that other products can replace products with a poor turnover. Another advantage is the generally high revenue potential.

On the other hand, companies focusing on transactions face difficulties when it comes to entertainment. It won't be easy to offer really entertaining content when sufficient transaction revenues are required. This leads to another disadvantage of the transaction model. Although some flexibility can be seen in the choice of products, this model is in general less flexible, because it is geared to transaction-based revenues. Extending operations in order to obtain additional revenues in the Get Audience or the Sell Audience area will be difficult.

Our perception of today's broadband media industry landscape is characterised by most of the companies relying on one of the business models mentioned above. They are all facing the danger of getting stuck in the middle. Looking forward, we see two possible developments. Some companies abandon additional revenues that do not focus on their core business. They have learnt that successful integration of new activities is costly and time-consuming. It requires management commitment and cannot be realised without significant and consistent efforts. These companies concentrate on their core competencies and align their operations towards a focused business model.

In contrast, others are proceeding to integrate all possible revenue streams. They need to actively push this strategy by aligning their strategic intent to this diversification. In order to implement this strategy, it is necessary to build up new skills, capabilities, networks and strategic partnerships with significant investments. This approach leads to a balanced business model.

The Focused Model

Get
Audience

Example : Kinowelt Medien AG

A recent example of the focused model is Kinowelt Medien AG. After facing difficulties with the business divisions of merchandising and several Internet activities, the company has now decided to fully concentrate on the more profitable area of content production. Its activities include various licensing, cinema film distribution and DVD production, thereby reflecting the general idea of using content across various media. Another example of the focused model is provided by content distributors who rely on their "bandwidth for money" business model, aiming mainly to achieve economies of scale by trying to gain high market shares, for example all over Europe.

The focused model exclusively operates in one area. It follows the strategy of concentrating on the company's core competencies. The main advantage of this model is the possibility to leverage the company's internal resources and capabilities as effectively as possible. It is an interesting alternative, especially for content producers, because their core business is expected to be a profitable one.

The inherent disadvantage of this type of business model is the high risk factor resulting from complete dependence on one single market.

In general, focused business models fully concentrate on revenues from what we call the Get Audience. However, it is also possible to concentrate on Sell Audience or transaction revenues. One example is companies specialising on interactive advertising production.

The Balanced Model

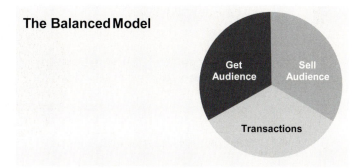

Finally, the balanced model tries to generate a substantial proportion of revenues in every basic area. In terms of the media business mechanics, this means a special kind of diversification. The approach has two great advantages. Firstly, it is the only way to exploit the full revenue potential. The company tries to maximise all revenue streams simultaneously, without omitting single opportunities. Secondly, the balanced model is an attempt to minimise risk. This reduces dependence on single markets, today a problem for many companies depending on the advertising market, especially in the Internet realm.

Among the disadvantages, we see on the one hand the required ability to operate efficiently in all areas. With respect to the large portfolio, management issues will be more complex and difficult than in other models, where a certain degree of specialisation takes place. On the other hand, problems may arise in maintaining a clear identity. What does the company stand for? How can the brand be positioned consistently? Bearing in mind that a strong brand is one of the key success factors, this issue has to be taken seriously.

Taking the idea of exploiting as many revenue streams as possible, we consider the described balanced model to be a particularly promising type of future business model for interactive broadband services. But all of the business models shown are combinations of revenue streams. Revenue forms the basis of every business model, which is why this was taken as the main indicator to show the conception of such models. But as we have learned in the previous sections, business modelling is more than just combining revenue streams. It is about value proposition, customer knowledge, core competencies and others, too. Hence, in the end, the challenge is to find an intelligent combination of all these aspects depending on the chosen strategic intent and appropriate to the industry's and company's context to build a sustainable position in the competition.

5.4 Outlook: Interactive Broadband Media Scenarios

Thinking about the question of where the whole interactive broadband service industry is heading, we found several important trends throughout our study:

- Concerning content, evidence was found for a trend towards offering value-added, attractive content and services combined with a stronger emphasis on customer education. This will increase the customers' willingness to pay for content and services and supports the future importance of Get Audience revenues.

- Secondly, the pacemaker function of technology determines the development of any content and service. Especially after the breakdown of the new economy, we see the trend that companies are becoming risk averse and don't want to get ahead of the technology curve. The diminishing interest of major companies to invest in small start-up companies that consist mainly of innovative ideas is a corresponding observation.

- However, a technological revolution is set to take place and will be supplemented in terms of sophisticated billing solutions, new aggregation engines, further development of digital content services and online sales engines. This revolution allows true and high-value interactivity forms - the basis for a new quality of transactions. This will lead to significant growth in transaction revenues, but also to an increase in paid content and services.

- The fourth trend is the growing perfection used to fully exploit the various Sell Audience revenues. Customer insight, permission advertising, all forms of merchandising and innovative event- and intention-centred marketing are developed in addition to traditional advertising, leading to additional revenues. This trend is in line with the general shift from the Get Audience to the Sell Audience that has been observed for many years.

- Fifthly, the experience throughout Europe shows that interactive broadband media in most cases is a local business, influenced by different cultures, legal aspects and infrastructure developments. Customer adoption behaviour, in particular, differs between countries. Hence, only some business models can be spread throughout all countries. This leads to the current trend of thinking globally and acting locally. With the further development of the local markets, this will result in consolidation, especially triggered by the acquisition of local companies by some big multi-nationals.

- Finally, there is a development towards multi-channel access. As soon as the necessary technological basis is available, we will see the advent of an increasing number of MCAPs. As a result, this market will consolidate quickly because of brand issues.

Guided by these trends and an analysis of their influence on the interactive broadband media industry landscape, a prediction of the future developments in this industry has been created. To illustrate our findings (see Chart 46), the future scenarios are organised according to two main dimensions. On the vertical axis, the macro-perspective is taken into account. The top end represents a fragmented market, characterised by many small competing companies. The bottom end shows a consolidated market, with only a small number of large companies dominating. On the horizontal axis, we consider the micro-perspective, with a range of differently aligned business models of the individual market players. On the right hand side, we see companies focusing on specific activities and revenues, with the discussed focused model as the extreme, whereas on the left hand side, companies are trying to diversify their operations and revenue sources, with the discussed balanced model as the most consequent form.

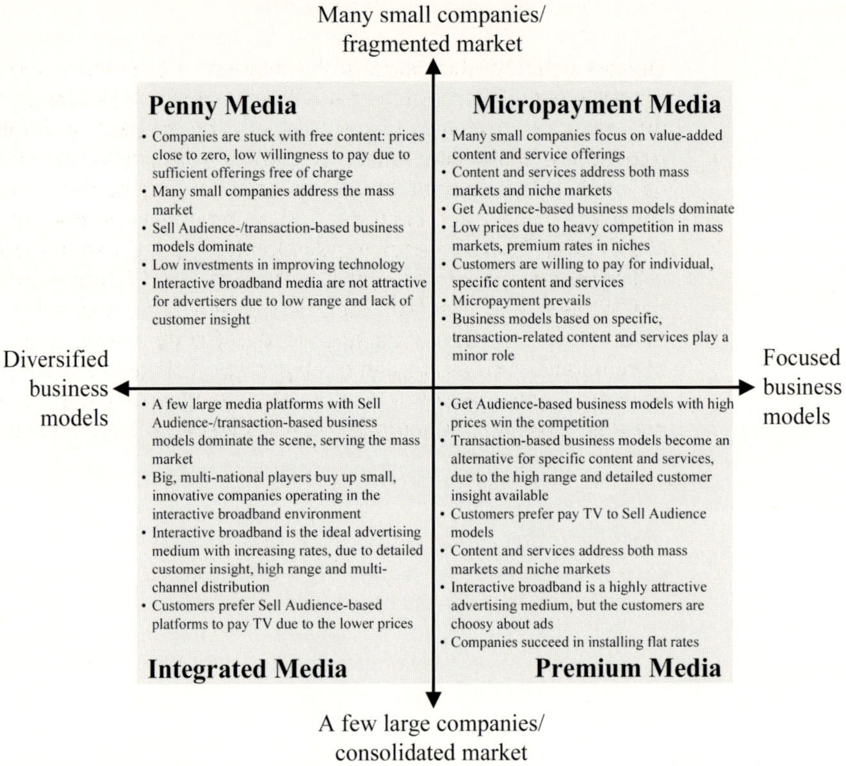

Many small companies/
fragmented market

Penny Media

- Companies are stuck with free content: prices close to zero, low willingness to pay due to sufficient offerings free of charge
- Many small companies address the mass market
- Sell Audience-/transaction-based business models dominate
- Low investments in improving technology
- Interactive broadband media are not attractive for advertisers due to low range and lack of customer insight

Micropayment Media

- Many small companies focus on value-added content and service offerings
- Content and services address both mass markets and niche markets
- Get Audience-based business models dominate
- Low prices due to heavy competition in mass markets, premium rates in niches
- Customers are willing to pay for individual, specific content and services
- Micropayment prevails
- Business models based on specific, transaction-related content and services play a minor role

Diversified business models ←→ Focused business models

Integrated Media

- A few large media platforms with Sell Audience-/transaction-based business models dominate the scene, serving the mass market
- Big, multi-national players buy up small, innovative companies operating in the interactive broadband environment
- Interactive broadband is the ideal advertising medium with increasing rates, due to detailed customer insight, high range and multi-channel distribution
- Customers prefer Sell Audience-based platforms to pay TV due to the lower prices

Premium Media

- Get Audience-based business models with high prices win the competition
- Transaction-based business models become an alternative for specific content and services, due to the high range and detailed customer insight available
- Customers prefer pay TV to Sell Audience models
- Content and services address both mass markets and niche markets
- Interactive broadband is a highly attractive advertising medium, but the customers are choosy about ads
- Companies succeed in installing flat rates

A few large companies/
consolidated market

Source: Accenture Analysis

Chart 46: Interactive Broadband Media Scenarios

Based on the results of this point of view, four resulting market scenarios can be found:

- Penny Media,
- Micropayment Media,
- Integrated Media and
- Premium Media.

Their characteristics are indicated in Chart 46.

Our prediction of the future of the interactive broadband media industry is depicted in Chart 47. The hatched area stands for the situation as we see it today. The market is highly fragmented with many small companies and only a few multinationals. Most companies rely on either a Sell Audience, Get Audience or transaction model with an emphasis on the Penny Media area. This tendency is reflected by the end-users' limited willingness to pay. Initial steps in the direction of consolidation can already be observed, especially in the content distribution area. However, the emphasis of the Penny Media area is obvious and goes along with our hypothesis of shifting revenues from the Get to the Sell Audience. This development pushes the pendulum further in the area of Sell Audience and transaction.

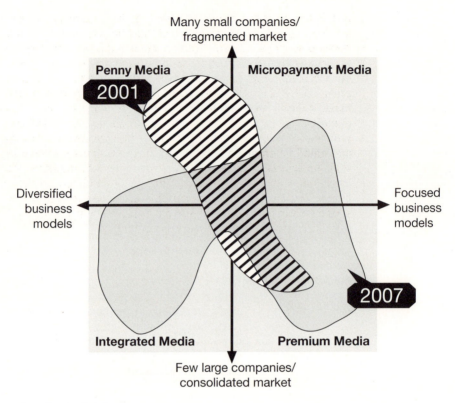

Chart 47: Future Development of the Interactive Broadband Media Industry

The grey area in Chart 47 is our prognosis of the situation in 2007. Market consolidation is expected as a result of shake-outs and further mergers and acquisitions. Most players will have aligned their operations either towards a diversified business model or towards a more focused business model. We expect that this will cause the pendulum to swing back slightly in the direction of Get Audience revenues. End customer willingness to pay will increase because of further customer education accompanied by increasing demand on the one hand, and joint efforts of the remaining companies on the other hand. The Premium Media area will mainly be occupied by content producers and distributors, whereas Integrated Media is seen as a home primarily for content aggregators.

In addition to these expected changes, interactive broadband service players must pay careful attention to further future developments yet unclear. We expect the most significant topics to include the ongoing end-user device competition. Will there be a breakthrough for set-top boxes deployed by the access providers? Will digital video recorder penetration reach a critical mass, thereby altering interactive service providers' business models by giving the end customer more control over traditional broadcasting? Is the trend moving towards television sets with integrated Internet services or towards PC-based solutions? And, finally, what does the actual development imply for the individual business models? The customer-driven decision concerning the preferred devices has the power to change the scene, forcing interactive broadband service players to align their business models to fit the new context.

6   Sources

Accenture Publications

"Beyond the ultimate device: From "anytime, anywhere" to "always and everywhere"", Anatole V. Gershman, Outlook-Point of View, 2000-2001

"Bigger than the Internet: The implications of digital television for financial service companies", Janar Weerasingam, Gary Fitzgerald, Paul Greenhalgh, Outlook-Point of View, 2000-2001

"Building Future Markets - Igniting the Broadband Revolution", Accenture Network Service Line, 2001

"Customer relationship management investments finally ring true", Betty A. McCabe, Mark T. Wolfe, Outlook-Point of View, 2000-2001

"Fault lines in customer management", Rainer Famulla, Outlook-Point of View, 2000-2001

"Getting your money's worth: Focusing chemical companies' customer relationship management investments", John Aalbregts, Stephen Dull, Outlook-Point of View, 2000-2001

"How do they know their customers so well?", Thomas Davenport, Jeanne G. Harris, Ajay Kohli, Accenture Institute for Strategic Change, August 2000

"Internet privacy: A look under the covers", Paul F. Nunez, Ajit Kamil, Outlook-Point of View, 2001

"Looking throu the customer's lens", Bill Crothers, Philip Tamminga, Outlook-Point of View, 2000-2001

"Measuring the payback on customer relationship management", Stephan F. Dull, Outlook-Point of View, 2000-2001

"Models of Customer Value", Jeanne G. Harris, Ajay Kohli, Accenture Institute for Strategic Change, July 2000

"Moving Customers - Building Customer Relationships and Generating Demand in the Mobile Data World", Tunc Yorulmaz, Brochure, April 2001

"New media. New value: The music revolution", James Anderson, Perzon Mody, Outlook-Point of View, 2000-2001

"Reinventing Cable-TV Business ", Thomas Herbst, Gerhard P. Thomas, Nikolaus Mohr, Christian Sciullo, Accenture Study, 2000

"Riding the Next Wave: The Strategic Implications of Interactive Digital TV and Broadband Services in Asia Pacific", Stephen C. Snyder, Accenture Study, April 2001

"Turning customer transaction data into business results", Jeanne G. Harris, Outlook-Point of View, 2000-2001

"Unlocking value through customer insight", Rainer Famulla, Bernhard Klein-Wassink, Outlook-Point of View, 2001

"What makes a good business model anyway? Can yours stand the test of change?", Jane Linder, Susan Cantrell, Outlook-Point of View, 2000-2001

Annual Reports

Bertelsmann, 1999/2000

KirchMedia, 1999

KPN, 2000

ProSiebenSat.1 Media AG, Interim Report 2000 (six months)

Vodafone AirTouch Plc., Interim Report 2000 (six months)

Books

"Bundesdatenschutzgesetz", Bundesbeauftragter für den Datenschutz, 1998

"Der Verlagskaufmann", Reinhard Mundhenke, Marita Teuber, Societäts-Verlag, 1998

"Digitales Fernsehen in Deutschland", Hans Paukens, Edition Grimme, 2000

"Einführung in das Datenschutzrecht", Marie-Therese Tinnefeld, Eugen Ehmann, R. Oldenbourg Verlag, 1997

"Marketing", Kotler & Blieme, Schäffer-Poeschel Verlag, 2001

"Merchandising", Karin Böll, Verlag Reinhard Fischer, 1996

"Merchandising und Licensing", Karin Böll, Verlag Vahlen, 1999

"Permission Marketing", Seth Godin, Simon & Schuster, 1999

"Residential Broadband", George Abe, Cisco Press, 2000

"Unleashing the Ideavirus", Seth Godin, Do You Zoom, Inc. Internet publishing, 2000

"Werbung im Pay-TV", Hans Paukens, Edition Grimme, 2000

Investment and Market Reports

"Ad Sales Strategies", Jupiter, April 1998

"Asset Transformation", Mindshare, 3rd Quarter 1999

"Broadband Access: New Business Models", Ovum, 2000

"Broadband Commerce", Jupiter, 2000

"Broadband consumer Internet access markets in Europe, 1999 - 2004", Datamonitor, June 2000

"Broadband Content splits", Forrester, October 2000

"Broadcast and Cable", Jupiter, June, 1998

"BSkyB", Dresdner Kleinwort Benson, April 6, 1999

"Commerce site metrics", Jupiter, 2000

"Consumer Content Strategies", Jupiter, June 21, 1999

"Consumer Market Trends: Germany", Gartner Group, June 9, 2000

"Consumers' buying intentions for PC broadband Internet access", Gartner Group, September 1999

"Consumers' use of Digital TV", Oftel, July 2000

"Convergence: The Big Themes", Paul Kagen Associates, 2001

"Copyright and Intellectual Property", Jupiter, June 1999

"Digital Home in Europe: Perspective 2003", Datamonitor, February 2000

"Digital Home Industry Overview", Banc of America Securities, September 2000

"Digital Rights Management", Jupiter, November 29, 2000

"Digital TV & Telecoms: Opportunities and Threats for Market Players", Ovum, 2000

"Digital TV markets in Europe, 1999 - 2004", Datamonitor, March 2000

"Distributing Digital Assets", Jupiter, March 8, 2000

"Dynamic Merchandising", Jupiter, 2000

"eBusiness Infrastructure", Billinmon.com, April 2000

"e-Home: Consumer Market Germany", Gartner Group, 1998

"E-mail Marketing", Jupiter, November 3, 2000

"European Digital TV", Jupiter, 2000

"European Media Companies face the global Internet", Jupiter, 2000

"European Media Companies face the global Internet", Jupiter, September 2000

"European Online Demographics", Jupiter, 2000

"European Pay TV and Cable 2000", Merrill Lynch, Decemer 2000

"From Napster to Snapster", Jupiter, October 4, 2000

"Global iTV Markets", Jupiter, 2000

"Inflection Technologies", Jupiter, 1999

"Interactive Adverising: New Revenue Streams for fixed and Mobile Operators", Ovum, 2000

"Interactive Advertising on Post-PC Platforms", Jupiter, 2001

"Interactive Television – Coming Soon to a Screen Near You", Marc Adams, Parul Anand, Sebastien Fox, Kellogg Tech Ventures, 2001

"Interactive Television (iTV)", ING Barings, September, 2000

"Interactive TV", Hoak Breeedlove Wesneski & Co., December 8, 2000

"Interactive TV Cash Flows", Forrester, August 1999

"Is the channel dead - The impact of interactivity on the TV industry", Datamonitor, February 2001

"ITV Advertising Landscape Emerges", Jupiter, September 20, 2000

"ITV Portals", Jupiter, 2000

"Licensing and Syndication", Jupiter, 1999

"Managing Content Assets", Jupiter, 2000

"Media: Cable and Entertainment", Morgan Stanley Dean Witter, January 2, 2001

"Micromarketing", Jupiter, 2000

"Mobile Content and Applications", Jupiter, 2001

"Networked Personalization", Jupiter, January 10, 2001

"NTT DoCoMo - The i-Mode Story", Goldman Sachs, September 20, 1999

"Online Shopping in Germany", Jupiter, 2000

"Paid Content ", Jupiter, August, 1999

"Proactive Personalization", Jupiter, 1999

"Prospects for Digital TV in Europe", Datamonitor, 2000

"Regional Broadband Networks: Germany", Ovum, February 2001

"Streaming Audio Advertising", Jupiter, 2001

"Television Home Shopping Industy Forecast", Morgan Stanley Dean Witter, January 2, 2001

"Television Online Landscape", Jupiter, 1999

"Television: Interactive TV Projections", Jupiter, 2000

"The future of IP Telephony 1999 - 2000", Datamonitor, 1999

"The Media Handbook - January 2000", Merrill Lynch, 2000

"The Online Advertising Report", Jupiter, September 1999

"The Race to build the Broadband United Kingdom", Goldman Sachs, August 12, 1999

"The Technology and Internet Primer", Morgan Stanley Dean Witter, Decemer 2000

"Trends and Outlook", Jupiter, 1999

"Valuing Entertainment Rights", Jupiter, 2000

Newspapers and Magazines

Businessweek

Computerwoche

Die Welt

Financial Times Deutschland

Focus

Frankfurter Allgemeine Zeitung

Freizeit aktuell

Handelsblatt

Mastering Management

Net-Business

Neue Züricher Zeitung

Produkte und Technologien

Süddeutsche Zeitung

The NewYork Times

The Wall Street Journal

Wirtschaftswoche

Werben und Verkaufen

Websites

absatzwirtschaft.de

aoltimewarner.com

ard-digital.de

bertelsmann.de

bigbrother.de

brainpool.de

constantin-film.de

contentguard.com

contentworld.com

ddv.de

diepresse.at

digitv.de

education.com

em-tv.de

hot.de

intertrust.com

jup.com

kirchmedia.de

marthastewart.com

mobile.one.at

mtv-media.de

napster.com

networkadvertising.org

nttdocomo.com

n-tv.de

onvista.de

premiereworld.de

qvc.de

reuters.com

rtlgroup.com

spiegel.de

streamgate.de

telefonica.es

telekom.at

terralycos.com

t-motion.de

ufa.de

vivendi.com

vizzavi.net

web.de

webtv.com

wuv.de

yahoo.com

Workshops, Presentations and Points of View

"CRM: The changing economics of customer relationship", Cap Gemini and International Data Corporation, May 1999

"E-Business Opportunities for German TV Companies", PriceWaterhouseCoopers, June 1999

Interactive TV & Advertising: "Was morgen anders ist kann heute eine Chance sein", IQPC, Berlin, 19./20.02.2001

Interactive & Internet TV: "Konvergenz von Internet und Fernsehen", Marcus Evans, Cologne, 24./25.10.2000

Interactive & Internet TV II: "Die Verbindung von Fernsehen und Internet wird die bisherige Medienwelt nachhaltig verändern", Marcus Evans, Hamburg, 04./05.04.2001

"NetResults 2000", Arthur Andersen, 2000

"Towards consumer marketing in the new millennium", NCR, 1998

"Trust and privacy online: Why Americans want to rewrite the rules", The Pew Internet & American Life Project, August 2000

"Vernetzung von Internet und TV", MTV Europe, October 2000

Index

171

Abbreviations

3G	third generation
ADSL	asymmetric digital subscriber line
ARD	Arbeitsgemeinschaft der öffentlich-rechtlichen Rundfunkanstalten der Bundesrepublik Deutschland
ASG	Austria, Switzerland, Germany
ASP	application service provider
CATV	cable TV
CPM	cost per thousand impressions
CRM	customer relationship management
DAM	digital asset management
DCS	digital content services
DRM	digital rights management
DSL	digital subscriber line
DT	digital terrestrial
dTV	digital TV
EDGE	enhanced data rates for GSM evolution
EPG	electronic program guide
ERP	enterprise resource planning
ETL	extract/transform/load
FTD	Financial Times Deutschland
FTTC	fibre to the curb
FTTCab	fibre to the cabinet
FTTH	fibre to the home
FTTR	fibre to the remote
GPRS	general packet radio service
GPS	global positioning system
GSM	global systems for mobile communications
HDSL	high bit-rate digital subscriber line
HSCSD	high speed circuit switched data

idTV	interactive digital TV
IP	Internet protocol
ISP	Internet service provider
LEO	low earth orbits
MCAP	multi-channel access provider
MEO	medium earth orbits
PDA	personal digital assistant
PLC	power line communications
PPV	pay per view
SDSL	symmetric digital subscriber line
SME	small and medium enterprises
SMS	short message service
SoHos	small offices, home offices
TDMA	time division multiple access
UDSL	universal digital subscriber line
UMTS	universal mobile telecommunications system
USP	unique selling position
VCR	video cassette recorder
VDSL	very high bit-rate digital subscriber line
VoD	video on demand
VoIP	voice over IP
VPN	virtual private network
VSAT	very small aperture terminal
WAP	wireless application protocol
WLL	wireless local loop
xDSL	digital subscriber line

Dr. Mohr, born in 1962, is a Senior Manager for strategy consultancy at Accenture's Frankfurt office and has spent the last few years concentrating on the area of telecommunications & media entertainment. He has nine years of consulting experience. In addition to organisational and strategic issues he focuses on advising companies as they enter the market in the convergence area of "interactive broadband media". Dr. Mohr has been in charge of several surveys in this area and is the author of several management books as well as various publications in management magazines. He has a master's degree in business administration, as well as a PhD in business studies.

Gerhard Thomas, born in 1957, can look back on 13 years of project and consultancy work in the areas of telecommunications, and public and private service provision. He has worked for Accenture since 1995. In his function as a partner, Gerhard Thomas deals with strategic issues in the world of telecommunications and the media and advises companies on establishing and expanding their business activities. He is responsible for projects in the areas of optimising customer operations and enterprise management solutions. Mr. Thomas has particular specialist knowledge in designing and implementing customer information and settlement systems.

Accenture is the world's leading provider of management and technology consulting services and solutions, with more than 70,000 people in 46 countries delivering a wide range of specialised capabilities and solutions to clients across all industries. Accenture operates globally with one common brand and business model designed to enable the company to serve its clients on a consistent basis around the world. Under its strategy, Accenture is building a network of businesses to meet the full range of any organisation's needs - consulting, technology, outsourcing, alliances and venture capital.

Accenture's Communications & High Tech Market Unit serves the rapidly converging areas of Communications, Electronics & High Tech, and Media & Entertainment. With 14,000 dedicated professionals the unit provides services to more than 80% of the world's largest electronics, communications, media and entertainment companies, as well as leading "dot coms".

For more information: http://www.accenture.de, http://www.accenture.com.

Accenture GmbH
Otto-Volger-Straße 15
65843 Sulzbach/Taunus
Germany
Phone: +49-6196-57 66047
Fax: +49-6196-57 66433

Dr. Nikolaus Mohr
E-mail: nikolaus.mohr@accenture.com

Gerhard P. Thomas
E-mail: gerhard.p.thomas@accenture.com